CSR, Sustainability, Ethics & Governance

Series editors
Samuel O. Idowu
London Metropolitan University, London, United Kingdom

René Schmidpeter
Cologne Business School, Cologne, Germany

More information about this series at http://www.springer.com/series/11565

Thomas Osburg • Christiane Lohrmann
Editors

Sustainability in a Digital World

New Opportunities Through
New Technologies

 Springer

Editors
Thomas Osburg
Fresenius Business School
Munich, Germany

Christiane Lohrmann
FranklinCovey Leadership Institute GmbH
Grünwald, Germany

ISSN 2196-7075   ISSN 2196-7083   (electronic)
CSR, Sustainability, Ethics & Governance
ISBN 978-3-319-54602-5   ISBN 978-3-319-54603-2   (eBook)
DOI 10.1007/978-3-319-54603-2

Library of Congress Control Number: 2017940417

Cover illustration: eStudio Calamar, Berlin/Figueres

Printed on acid-free paper

This Springer imprint is published by Springer Nature
The registered company is Springer International Publishing AG
The registered company address is: Gewerbestrasse 11, 6330 Cham, Switzerland

Foreword

Nowadays, digitalization influences nearly every aspect of our life. The seemingly endless global flow of digital information has revolutionized not only our economy by creating manifold application opportunities. In fact, the Internet of Things, Big Data, and digital innovations embody a megatrend. While this development offers various intriguing opportunities, it also includes manifold serious challenges. Besides data security and property rights, one of the biggest questions to answer is whether we can shape a *sustainable* digitalization.

A sustainable development of all societies is of crucial importance for the future of our planet. The United Nations estimate that by 2050 our planet will be home to more than nine billion people. This tremendous demographic change will certainly have a profound impact on our Earth. Humankind has already transformed about half of our planet's land surface, and the oceans, too, are in a much worse state than they were just a few decades ago. Will this development eventually exceed our planetary boundaries? The Sustainable Development Goals (SDGs) attempt to curb an unbearably negative anthropogenic effect on the planet. They define a developmental corridor and a welfare concept with which a considerable increase in global population might be made tolerable. In my opinion, the SDGs are a highly promising instrument as they apply to all states and not just developing countries.

Whether the megatrend of digitalization will contribute toward a sustainable development in the long run is dependent on how we shape it. I will briefly highlight key challenges alongside the pillars of sustainability to pinpoint vital and promising fields of policy activity.

Digitalization is in need of vast quantities of energy, for example, to power data centers. Power usage of these centers alone amounts up to 2% of the global energy demand. The concept of Green Computing attempts to reduce the environmental impact of IT hardware, especially by decreasing its energy consumption—and hence carbon emissions output—in numerous ways: It can for example contribute to achieving a higher degree of capacity utilization of servers or more energy-efficient cooling systems in data centers. Another key aspect is resource efficiency and recycling. IT devices often contain dozens of different (rare-earth) elements whose extraction damages soils, groundwater, and wildlife. Better design and production standards can result in less dependency on raw materials. Our long-term goal has to be a completely closed resource cycle. A pressing challenge in that regard is how to avoid rebound effects based on the increase in energy and resource efficiency.

From an economic point of view, it is clear that we need to tackle two challenges. First, we need to encourage companies to implement transparent and sustainable supply chains. The global market economy in turn has to reward companies that live up to their entrepreneurial responsibility. The German Federal Government sets best-practice examples by initiating multistakeholder initiatives in numerous industrial sectors along the whole supply chain, for instance, the Partnership for Sustainable Textiles. Second, public organizations need to better steer their influence in public procurement toward sustainable standards, for instance, by demanding sustainability certificates for an award of contract.

A key principle of sustainability thinking is the idea of sharing knowledge. In order to enable every human being to acquire knowledge, open Internet and data access are necessary. This groundwork allows for a transformative education dynamic, redirecting societies toward sustainable development. The UNESCO Global Action Programme on Education for Sustainable Development stresses this important feature. Digitalization can reinforce the positive educational effect of this approach, for instance, through e-learning platforms. These aspects of the social dimension of sustainability are immensely important as they strengthen education and learning on all levels, enabling future generations to meet their own needs.

A sustainable digitalization is possible. However, there are many challenges ahead that we need to tackle actively and comprehensively. The scientific community already plays a pivotal role in providing alternatives and we should continue to encourage research in this very important field.

I highly welcome this book and look forward to gaining profound insights into the compatibility of digitalization and sustainable development.

Berlin, Germany   Andreas Jung
April 2017

Preface

Digitalization of all areas of life brings dramatic changes to our societies, our democracies. It has started quite a long time ago already and is taking place in a big way right now. It is challenging our fundamental values, constitutional principles, and legal environment. The necessary answers have not yet been given. We're just at the beginning of a deep transformation to a digitalized society. It comes together with the ongoing globalization and an individualization which drives the economic developments across the globe. This is a revolution of our life which brings as many chances as dangers to all of us. It is therefore important and urgent to talk about the challenges and consequences of each aspect in this transformation. Sustainability is a very relevant one. This book collects different useful perspectives on sustainability in the digital world. Looking at governance, mobility, production, work life, and corporate responsibility, it gathers numerous relevant areas which will be subject to changes and new models through digitalization. It will be key to deepen this exchange. Since some years also policymakers debate the specific relevance of digitalization to all areas of life. Slowly we're realizing the dramatic extent in which all this will take place. Every new innovation and every change to today's life could have heavy consequences on other areas and on the balance in society, economy, or environment. It is therefore absolutely necessary to implement sustainability already in the design of new innovations and developments.

One example to achieve sustainable developments in the digital world is to get away from fragmented regulation on the digital market. The path toward a Digital Single Market in the biggest common market—the European Union—is a major step toward consistent application of legal principles and rules. In particular, the creation of unified frameworks in the field of data protection and telecommunication standards has been historic changes toward a sustainable digital environment. The EU is not only giving an answer to the cross-border nature of the digital world by saying goodbye to national competences and differences but also building a pillar for future global standards, which will need to be discussed sooner than later. It has been hard work to overcome national differences, but it will be even harder to continue these developments without losing legitimacy or democratic principles. No matter where we will end up at the end of this deep transformation, we will need to assure that we're not sacrificing the civilizing achievements and in particular human dignity and self-determination in the liberal societies we built over the last decades and centuries. It will be therefore imperative to look very carefully on the side effects of every new step into the digital world of tomorrow, especially with regard to the sustainability for mankind and environment.

Munich, Germany
April 2017

Jan Philipp Albrecht

Preface

We experience a world in transformation. Pundits who not so long ago claimed either the end of history or the advent of a time with no geography are being challenged by everyday news. We live in an era in which knowledge is being produced and made accessible through a variety of means, yet, rather than experiencing full control we tend to feel less certain. For example, companies define the environment as a VUCA one. Volatility, uncertainty, complexity, and ambiguity are some of the features of the world we live in.

Never have strategists been so accurate in their use of the dichotomy. This is a world in which opportunities and threats abound. On the one hand, the world might have well become a hotspot, a nest full of challenges and tensions. Artist Mona Hatoum builds a large sphere in which continents are profiled in bright red neon. The viewer confronts something like a large round structure with intense red light coming out of it. No corner of the globe, we are reminded daily by media, lives unconnected and unchallenged. The red shows the tension and the urgency.

At the same time, the world has never been more conscious of the global challenges we experience. Global warming and the eventual climate change it provokes may be the just the epitome of such consciousness, but if we consider for a moment the impact of digitalization this might well just be one of the major changes we are about to undergo. Consider for a moment the economics of it.

Large manufacturers are pondering on how business models and products are radically changing. And they do it fast. London cabs observe today how Uber drivers take a big slice of a business that was assumed to be anchored in rock. At the same time, Uber is a company which does not fit the usual management structure. No managers in view and no headquarters on the map make the company a virtual one. But the bus does not stop here, Uber drivers may well be expendable and once systems become more safe. Then, who would be in need of a driver? Cars might well do the job without close supervision.

What then for job security, for long-term business plans, or for the kind of variables that economists used to consider as necessary and sufficient variables, such as labor and capital? That is, what about a world in which jobs are becoming a precious asset. Luddites will lose the battle again; still, societies are made of people who experience common problems and get together to solve them. These are not issues that can be simply left to mutual adjustment, and, as anthropologist David Graeber has rightly pointed out, bonds came first and self-interest later.

This is a book which looks at the world's ambition to provide meaning to a term, sustainability, that has captured most of our hopes. How can we make a world more sustainable, that is, how can we be sure that our current investments in education and business are the right ones to bring long-term well-being. How can we make the most of the transformations we experience and make them work in our favor. There will be multiple avenues, and this book will rightly point to some of them.

Brussels, Barcelona   Alfons Sauquet Rovira
April 2017

Preface

Changes have been part of this world as long for as it has existed. However, the speed of changes our societies are experiencing is growing exponentially, not solely due to digitalization, but to a significant degree. It affects the ways we live and work, we learn and communicate, and also how we form options and live together.

Thus, fundamental values are challenged and are changing, with consequences for legal regulations and human principles. We are only at the beginning of learning about, let alone understanding, the consequences digitalization will have on the environment, on workplaces, and on education, to name just a few key areas. Combined with ongoing globalization and immense economic challenges, we need to find solutions that enable us to sustain not only the planet we live on but also the societies we live in. It will be a lot of work.

But, as there are two sides to every coin, and a glass can be half full or half empty, we also need to look at opportunities that a more digital world can bring. Think of dematerialization and how it can reduce burdens on the environment. Think of the tremendous possibilities in the healthcare sector and how they can help people across the globe, not only in healing, but also in the prevention of diseases in underdeveloped parts of our world.

As we are unable, and partially unwilling, to halt the progress, we need to focus on making technology a part of human lives. How can technology serve people and not the other way around? How do we make sure we don't leave parts of our societies behind? And how can we reestablish a trust in technology?

In few places is this more relevant than in education. As UNESCO states: "Education for Sustainable Development allows every human being to acquire the knowledge, skills, attitudes and values necessary to shape a sustainable future." And if you take a deeper look at the UN Sustainable Development Goals and specifically Goal 4 (Quality Education), it is clear to see that all the subgoals to this objective are only achievable with a certain level of digitalization that truly helps to make the lives of people better.

Education here has two challenges: to prepare the younger generations to design and implement a sustainable future, but also to use digital tools in meaningful ways to reach these goals. MOOCs, which already have the potential to reach many people in societies across the world (e.g., in rural areas), enabling them to participate in world-class education, are a promising example.

This book comes at a very timely moment and I highly welcome it. We are starting to understand sustainability from an ecological perspective, but need to learn to also understand its meaning for humans, in an increasingly digitalized world. The contributions in this book will help us to provide some much-needed answers.

Cologne, Germany   Tobias Engelsleben
April 2017

Introduction

Digital Innovations have become companions in our daily life. A lot of hopes and expectations go along with this development. Connected cars could save lives from road accidents, e-health could improve medical services for people, and smart technologies could cut carbon emissions. However, despite the potential innovations and possibilities of digital development, a lot of uncertainties and open questions remain. While a lot of focus has been on innovation and new technology as well as the "Green through IT" aspect, there was little discussion on what impact this has on supporting sustainable development of the society as a whole. Can digital innovations contribute to improve the quality of people's lives, achieve equitable growth, and help protect the environment? Do they help drive progress toward United Nations' recently formulated 17 Sustainable Development Goals (UN 2015)? This publication focuses on how digital development promotes at the same time commercial growth and sustainability issues.

This book goes beyond the existing "Green through IT" thinking that enriched the public debate for many years already. It does not only focus on technology and ecology only but includes the human perspective as it looks at how people benefit from the digital world in a variety of areas, like consumption, education, participation, and mobility. Furthermore, we look at how digital development challenges us in management and leadership as well as in ethics and responsibility. It is about the shift of perspectives toward sustainable society and world at large in the three fields of sustainability: social, environmental and economic.

Fig. 1 Triple bottom line of sustainable development (see Brundtland Report 1987)

Which developments do we witness? And: where are the blocks that prevent to realize digital development toward the SDGs to be fully unleashed?

The key questions of this book is: How does the ongoing move toward a digital world contribute—positive or negative—to a more sustainable world?

Call for a Critical View

As digital development moves on and becomes more and more present in everybody's lives, it is obvious that there are still many challenges to meet, hurdles to take, and obstacles to be overcome.

First of all, there are *political and regulatory challenges* related to the awareness of sustainable development as well as data analysis and security. While at the time new insights from data are needed, differences in regulatory requirements still slow the deployment of sensors and smart technologies. As a result, they increase the complexity associated and therefore add to their cost.

Secondly, we should also reconsider our *values and ethical concepts* and the notion of responsibility in the context of digital innovation. Many developments run ahead with little planning and control. Risks are underestimated due to the wish to grow. Therefore, we might also think about laws, regulations, or norms to organize the use of digital devices, especially in the context of addicability. Also, we need to rethink critically our responsibility and ethical approach about handling digital development in the context of sustainable development. Here, strategic management and leadership issues need to be addressed.

Finally, we will have to take a close look into the *possibilities and challenges of life in the digital world* as well as new opportunities regarding access to education and participation for development. Here, digitalization offers a lot of options as well as challenges and barriers at the same time. However, it will be shown that investment in these fields will need to increase in order to realize change in policy, company, and the environmental sector.

About This Book

This book connects several fields digital innovation through ICT with management, leadership, and ethical orientated thinking in the context of sustainable development. It aims to be inspiring, encouraging, as well as thought-provoking and critical at the same time. We brought together international thought leaders from academia as well as business and foundations to get diverse inspiring and thought-provoking input. Many of these perspectives call for a shift of paradigms. This book is for business leaders, academics, writers, and critics who care about life, value, and business development in a digital world and at the same time try to contribute to the values of a sustainable future.

Acknowledgements

The authors of this books are all renowned and leading experts in their particular field. All have been enormously cooperative in creating this book. We would like to give a special thanks to each of them for sharing their deep knowledge and personal ideas regarding digital development and thriving toward the SDGs. Furthermore, we would like to thank Christian Rauscher and his colleagues from Springer, who immediately supported the idea of the book and accompanied us with their profound knowledge and professional advice, which was very encouraging. It was also very satisfying to be supported by political representatives Jan Philipp Albrecht of EU Parliament and Andreas Jung of Parliamentary Advisory Committee for Sustainability of Deutscher Bundestag. A special thanks also goes to Prof. Alfons Sauquet Rovira and the ABIS Team for professional support. Also, we would like to thank Prof. Dr. Tobias Engelsleben, President of Fresenius Business School, as well as Prof. Dr. Martin Kreeb of the Fresenius Business School for supporting the academic approach of book.

Christiane Lohrmann
Prof. Dr. Thomas Osburg

Literature

United Nations (2016) Sustainability development knowledge platform. https://sustainabledevelopment.un.org/sdgs. Accessed 11 Nov 2016

United Nations (2014) The road to dignity by 2030: ending poverty, transforming all lives and protecting the planet. http://www.un.org/ga/search/view_doc.asp?symbol=A/69/700&Lang=E. Accessed 23 Oct 2016

Setting the Scene: The Relevance of the 17 SDGs for Digital Development

While digitalization rapidly changes our world, politicians and diplomats of all countries have agreed on a common political understanding of the common goals in a future sustainable world. In September 2015, the UN announced the 17 Sustainable Development Goals (SDGs) as a basis for the 2030 Agenda for Sustainable Development (UN 2015). Over the coming 15 years, member countries are urged to mobilize efforts to end all forms of poverty, fight inequalities, and tackle climate change while ensuring that no one is left behind. The UN developed the SDGs in cooperation with many stakeholders: NGOs and Foundation Representatives from Science, Politics, and Business.

Although the SDGs are not legally binding, governments are expected to take ownership and establish national frameworks for the achievement of the 17 Goals. Countries have the primary responsibility for follow-up and review of the progress made in implementing the Goals, which will require quality, accessible, and timely data collection. Regional follow-up and review will be based on national-level analyses and are urged to contribute to follow-up and review at the global level.

There are high expectations regarding digital developments to contribute to the SDGs such as improving people's lives: 1.6 billion people could benefit from more accessible, affordable, and better quality medical services through e-healthcare, while connected car solutions could save up to 720,000 lives annually and prevent up to 30 million traffic injuries. This helps ensure healthy lives and therefore could contribute to achieve SDG #3 (GeSi 2016). Also solutions for open education through the Internet, such as MOOCs are expected to increase education around the world. A possible contribution to SDG#4, which calls for inclusive and equitable quality education to promote lifelong learning opportunities for all. Moreover, in the environmental field it is called for a resilient infrastructure, to promote inclusive and sustainable industrialization and foster innovation. Therefore, solutions could enable greenhouse gas emissions reduction and drive market transformation for renewables, cutting carbon emissions by around 20% in 2030. A potential contribution to environmental protection is called for in SDG #13. In addition, we are facing challenges and opportunities in the markets including challenges and chances for producers, consumers, and stakeholders. Here, SDG # 9 becomes relevant: it calls for a resilient infrastructure, promotes inclusive and sustainable industrialization, and fosters innovation. Also SDG #12 which urges to ensure sustainable consumption and production patterns is to be considered. This only being a short summary of possible effects of digitalization on the SDGs, there are a lot more expectations in what digital innovation should and can contribute to sustainable development.

Despite all these optimistic viewpoints, we have to face and acknowledge several facts, laws, and regulations and other roadblocks on the way to a sustainable future matching with the UN SDGs. This publication gives insight into chances and possibilities in crucial areas of digital development in the context of a more sustainable world.

Prof. Dr. Thomas Osburg
Christiane Lohrmann

Literature

GeSI Report (2015) SystemTransformation: how digital solutions will drive progress towards the sustainable development goals, Accenture Strategy; www.systemtransformation-sdg.gesi.org. Accessed 23 Oct 2016

United Nations (2016) Sustainability development knowledge platform. https://sustainabledevelopment.un.org/sdgs. Accessed 11 Nov 2016

BASF (2016) UN sustainable development goals. https://basf.com/de/company/sustainability/employees-and-society/goals.html. Accessed 13 Nov 2016

Contents

Part II   Markets, Business and Stakeholders

Part III   Participation, Education and CSR

Part I
Governance, Strategy and Society

Sustainability in a Digital World Needs Trust

Thomas Osburg

1 Introduction

A farmer, who plants tomatoes and potatoes, can estimate the consequences of his work pretty easily—on the soil, on people, on the environment. But a trader, who uses millions of data from around the globe and acts based on algorithms, can't. We see a dramatic shift over the last three centuries when talking about sustainability.

When Hans Carl von Carlowitz first published his famous Sylvicultura oeconomica in 1713, the boundaries for acting sustainable were either the local forest or maybe an area that could be overseen rather easily. Not cutting more trees than what can regrow is easy to calculate and predictable, the consequences are rather clear and local (Sächsische Hans-Carl-von-Carlowitz-Gesellschaft 2015).

This well-understood concept of sustainability is now confronted with a technological leap we call Digital Revolution or Digital Transformation. Digitalization offers new possibilities and pathways of how to shape the future of living together. Predictive medicine enables the monitoring and curing of how infectious diseases spread globally. Algorithmic capacities allow for data processing and analysis that open up unseen capabilities. Digitalization bears consequences for transparency and accountability which create entirely new ways to shape, monitor, and govern sustainability. In conclusion, both megatrends, sustainability and digitalization, impose major transitions on our world and how we picture it. Ultimately, Digitalization will fundamentally change the structures of our societies (Müller von Blumencron 2016)

As the world is moving to *Digital*, more and more services are delivered online: Daily Papers, Banking, Education, Machines talk to each other and personal data is in some cloud. While a lot of focus has been on Innovation and New Technology,

T. Osburg (✉)
Fresenius Business School, Munich, Germany
e-mail: thomas.osburg@hs-fresenius.de

© Springer International Publishing AG 2017   3
T. Osburg, C. Lohrmann (eds.), *Sustainability in a Digital World*, CSR,
Sustainability, Ethics & Governance, DOI 10.1007/978-3-319-54603-2_1

there was little discussion on what impact all this has on supporting Sustainability goals (Osburg 2013). Why is that so? Typically, we view Digital products as *Carbon Light*, means that they are supposed to have little impact on emissions and pollution. However, producing and delivering digital products requires significant energy, produces carbon dioxide emissions and has significant impact on the society at large—how we behave, how we consume, how we work and how we live.

The Digital Economy offers enormous opportunities: You can reach rural and underdeveloped areas of this world with state-of-the-art education, intelligent machines can do jobs that humans don't want to do, patient care can improve through new forms of caretaking, and of course we all don't want to miss the comfort of accessing all our data anytime and anywhere.

One of the key questions in all this development is not yet fully answered: How does the ongoing move towards a digital world contribute—positive or negative—to a more sustainable world? Is a more digital world always more sustainable? What are the key focus areas to look at, what are the opportunities but also the challenges? How does the Society at large support all this?

In the wake of digitalization, megatrends such as mobile internet, the internet of things, big data or digital innovations are creating development opportunities faster than ever. Digital is a crucial driver for decent work, growth and well-being, and is having a profound impact across all sectors. The internet and digital technologies can and will boost economic, social and political development, including by vastly expanding the capacity of individuals to enjoy their right to freedom of speech and expression, which is key to empowering human rights (De Croo 2015).

But how much digitalization do we want? Do we always want more? Do we always need more? And do we even have a choice? How much are we ready to 'pay' for it, not in Euro or Dollars, but in potential loss of privacy and security. What tradeoffs do we need to make and what impacts will this have on society as a whole? What is the new role of Government? Protecting or enabling, i.e. like in Estonia that considers itself 'Country as a Service' (Domscheit-Berg 2016)?

2 Changing Concepts of Sustainability

Ecological Sustainability

Since the early 1980s environmental aspects of sustainability were the primary focus, with major concerns about air pollution and acid rain. Over time, discussions included other environmental aspects such as water and other natural resources, biodiversity, clean energy, agriculture and food. Now the theme of Climate Change is perhaps the predominant concern (Tardieu 2014). As such, Ecological Sustainability can be seen as the capacity of ecosystems to maintain their essential functions and processes in the long run.

The current emergence of Digital Solutions, Big Data, Internet of Things (IoT, connected objects and people) and so on will generate vast amounts of data that with the application of smart analytics and visualization techniques will help us to understand more about the way we interact with each other and our environment; with businesses; and with the world around us. Unlocking such insights will enable us to discover patterns for more sustainable behavior, for example (Tardieu 2014):

- Improving forecasts of natural events or disasters
- Optimizing global agricultural production and food supply
- Anticipating traffic congestion and managing low emission zones
- Limiting energy production up to the precise needs of consumers
- Allowing preventative maintenance that avoids failure and replacement

Even though it is mostly understood that delivering such a connected world and managing the resulting data will in itself impose an environmental load (i.e. server parks), the impact of digitalization on the environment, like virtualization, de-materialization, efficient hardware components, free air cooled data centers, etc. will certainly help to reduce negative environmental impact (GeSI 2015).

Economic Sustainability
Understanding Sustainability as a normative concept of 'capacity of ecosystems to maintain their essential functions and processes in the long run', economic sustainability is grounded in the use of various strategies for employing existing resources optimally so that a responsible and beneficial balance between business and society can be achieved over the longer term. It can be understood as the maximazation of revenue and profit while at the same time maintaining needed resources over a longer period of time (Osburg 2017). Within a business context, economic sustainability involves using the combined assets of the company efficiently to allow it to continue functioning profitability over time.

In addition, positive company behavior was partially encouraged by government policies that enabled positive financial impacts for those firms that engaged in sustainable activities (e.g. by subsidies) or penalized non-sustainable activities through taxation. This can be understood as the 'economization of environmental/ social aspects of sustainability' (Tardieu 2014). As a goal, there are tangible positive economic benefits to be expected from sustainable approaches to business, like

- Less waste, less energy consumption, time saved.
- Attracting consumers who are motivated by environmental concerns.
- Positive contribution to the Triple-Bottom-Line reporting of the firm.
- Using only needed resources through '... as a Service' concepts, enabled largely by Cloud Computing, where only the actual usage of a product or service is paid for.

Social Sustainability

The ability and willingness of a society to develop processes and structures that not only meet the needs of its current members but also support the ability of future generations to maintain a healthy community and intergenerational justice is a key component for concepts of Social Sustainability. At a larger scale, it also includes concepts of trust (to companies and institutions), ethical behavior (of organizations) and can be a parameter for the equitable distribution of a nation's wealth by providing people access to resources, goods and services to fulfill their needs.

Compared to ecological and economic aspects of Sustainability, where investors often benefit from its positive outcomes (i.e. minimization of energy consumption to help reduce costs and thus generating ROI and profit), those who invest in long-term technologies, solutions and polices for Social Sustainability are not likely to be those who will be able to benefit from them. 'The social aspect of sustainability thinking becomes a key success factor for our planets longer term future wellbeing' (Tardieu 2014).

This is an area where Digital Transformation will ultimately change the game, enabling new models of society often based on sharing—which is a key principle of sustainability thinking.

- New economic models, where providing personal data in exchange to free services or products.
- Ethical projects with usually little success in finding seed money can be facilitated through crowdfunding: trust is needed but sustainability is generally a world of trust.
- Mobility is obviously a way to enhance availability and connectivity, again building on trust of systems and people.
- The reinvention of work, often referred to as "Industry 4.0", is certainly a major breakthrough in delivering enhanced productivity, environmental benefits and collaborative work concepts.

Ultimately Social Sustainability can be understood as '...identifying and managing business impacts, both positive and negative, on people' (UN Global Compact 2016).

3   Digital Technology with Impact on Society

We live in exponential times. While the world has seen many dramatic changes over the years (Electricity, Industrialization, Trains, Information Technology, etc.), the speed of today's changes is the key challenge. At no time before in history people had so little time to adapt to societal and technology changes. While this brings tremendous progress, wealth and (sometimes) peace across the globe, we are now at a point in time where we need to realize that this might not be true for all.

There are people who will loose and people who will win. Maybe Charles Darwin (1859) was never so right as of today: 'It's not the strongest of the species that survive, nor the most intelligent, but the one most responsive to change'.

Translating these thoughts into today's world, we certainly do not talk about physical survival anymore, but economic win-lose situations. Digitalization will see a lot of winners, but, at least in the short to medium term, also people who are left behind. People, who are either unwilling or unable to follow today's societal and business development. These people are, at least today, not necessarily losing, but they are not winning and thus leaving (economic) advantages to the ones who are willing and capable to adopt (see Fig. 1).

We will see a divide into Digital Elites and Analog Illiterates with dramatic consequences for societies. And this leads to concepts of Social Sustainability. What kind of society do we want to have in the next years and decades? What impacts from Digitalization can we expect and how do we deal with it?

This section will deal with three major areas of digitalization, that will impact, more than others, Social Sustainability of our Societies.

- **Data**: A constant focus on data will be key in the coming years. What kind of data are available, who owns those data, how are those data used? What kind of acceptance is needed from consumers? And how can consumers keep their rights on their own data?
- **Algorithms**: What used to be a more technical term in the past is now quickly becoming a critical gate-keeper in today's information society. More and more decisions, at all levels, are determined by algorithms, which are a self-regulated sets of operational steps that need to be performed. But should algorithms

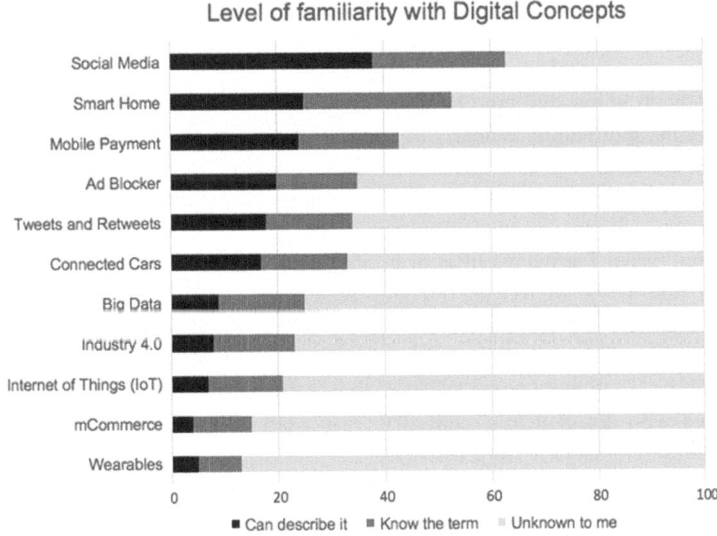

Fig. 1 Level of familiarity with digital concepts in January 2016 (TNS Infratest 2016)

determine our lives? Algorithms are always programed by humans, so what kind of credibility and social license to operate do these programmers and companies have?

- **Bots**: A more digital world, with digital processes all around us, will have significant impacts on our jobs in the future, as machines and (ro-)bots are increasingly capable to replace tasks people perform today. For the moment it remains unclear, if there will be positive or negative impacts, and within which timeframe. Will (ro-)bots take away our jobs? And, if so, which jobs? Current studies are partially contradictory.

Data

We all know by now that data is the new gold, the new oil or the new currency. Having access to customer data, for example, is key in remaining competitive in the next years and decades. We can see this battle for data amongst firms in nearly all business sectors: In the car industry, in retail, in the health sector and so on (Steinbrecher and Schumann 2015). The advantages are obvious, as the possession of data allows a much more detailed customer targeting and thus potentially better suited prices, offers and ad's (Wadhawan 2016). We call it 'personalized experience' and it is based on all the data collected from consumers at various levels and steps during the shopping process.

While this is a known phenomenon within the online world—Google estimates that it can look at more than 50 signals a person sends out while using a computer, i.e. location, browser, PC, pages visited, products bought, etc. (Pariser 2011a)—we are now seeing more and more in-store systems generating similar data and thus predicting the potential next moves of the shopper. Face-recognition software of cameras at the shop entrance are revealing if a known customer is in a good or bad mood, if he comes alone or with his wife, if he is dressed in business clothes or rather casual or if he belongs to the top-customers, so the manager will come immediately to greet him (Frey 2016). In-store tracking systems detect the paths any given customer takes, what products he is looking at for how long and what finally ends up in his shopping basket (Clauß 2013).

So staying anonymous is nearly not possible anymore for consumers who prefer not to be tracked or leave traces. All these tracking tools, however, have one commonality: While customers are increasingly aware that they pay with their data, somehow, it remains unclear to them, what kind of data they exactly hand out, and at what price. They have little or no influence and control of the usage of their own data, and they have to trust the data-owner that a confidential use of their data is guaranteed and is leading to advantages (more targeted ad's, free services, etc.) they might or might not use. This, however, is an unintentionally provided trust by the consumer. They were never asked, if they wanted this. 'It is not the countries or companies who perform a 'digital striptease', but citizens and consumers', concludes Daniel Domscheit-Berg (2016).

The World Economic Forum categorizes data revealed from consumers into 'Volunteered Data, Observed Data and Inferred Data' (World Economic Forum 2011). Some of the Observed data collections, as described above, will likely be around for the coming decades and mainly driven by available technology and regulatory frameworks. Contrary to all the Big Data discussions, which means the massive collection of Data for analytic and predictive use, there is an emerging discussion about the real value of all this Big Data. Lindstrom (2016) believes that there is still more value in constant observation of the consumers as a base for deducting the right conclusions from this observation. He calls it 'Small Data' and assumes that two out of three successful innovations stem from Small Data, not Big Data.

Where we will see major changes, however, is in the area of Volunteered Data. Today, this mainly centers around information revealed while subscribing to Newsletters or Customer Profiles, Bonus Cards and Social Networks. We will see an emergence of requests from companies to provide more (volunteered) data, that offer significant advantages to customers. For example, you might get a 10 Euro discount on a purchase, if you share a subset of personal data with the store, or a 10% deduction from your health insurance provider's bill if you prove a healthy and sustainable lifestyle, i.e. with the help of smart watches, step trackers, etc. (Rosenbach 2016).

Ultimately, it is nothing else than putting back the control of data value into the hands of the consumer, as he or she can now reveal personal data or not, it becomes an individual's choice. This could be seen as '. . .expanding the capacity of individuals to enjoy their right to freedom of speech and expression, which is key to empowering human rights' (De Croo 2015).

Algorithms

'Information technology is a formidable enabler of freedom. For example, it lowers barriers to freedom of expression and allows people to get a better grasp of their lives' (De Croo 2015). This statement reflects to a large degree a widespread thinking, about freedom of the Internet and same opportunities for all. But this is changing and these changes will have significant impact on many ways we live together. It touches upon the information people receive and that they use as a base for their behavior. Eli Pariser (2011b) calls it a 'Filter Bubble' and describes it as a result of web site personalization in which algorithms increasingly guess what a user would like to see and what he or she would not like to see. In a first step (see Fig. 2a), the user is still surrounded by a wide variety of media and opinions, but gets to see mainly the pieces of information within the inner 'bubble'. In step 2 (see Fig. 2b), the user is not even aware of anything outside the Bubble and has to assume that the world is only what is visible within this bubble. Different opinions or news are not reaching him or her anymore. The algorithms on the Internet act as

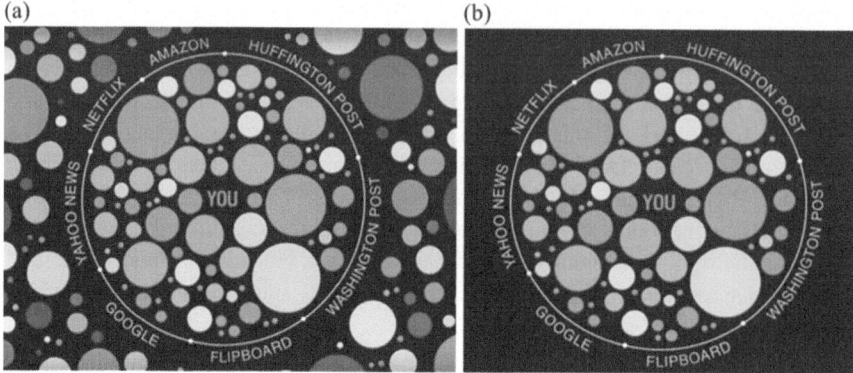

Fig. 2 Concept of the filter bubble (Pariser 2011a)

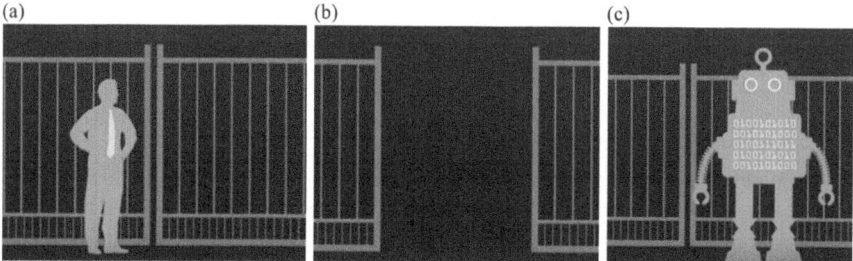

Fig. 3 Changes in "Internet Gatekeeping" (Pariser 2011a)

self-reinforcing forces to continuously reduce the breadth of information down to what the user might like.

The reason for this development is seen in the rise of the importance of algorithms. While there has always been a sort of control of information, usually through editors of journals, newspapers and TV emissions (see Fig. 3a), the first years of the Internet were dominated by openly available information for all (see Fig. 3b) with no or very little pre-defined content. Today, however, non-transparent algorithms have taken some kind of invisible control over what users see and read (Fig. 3c). Users get less exposure to conflicting viewpoints and are isolated intellectually in their own informational bubble. According to Pariser, the bubble effect may have negative implications for civic discussions (Pariser 2011b) and thus influence political elections and societal developments (Weingarten 2015).

Bots

'We hope that the current Industrial Revolution will develop as previous ones: Few jobs will disappear, but the power of Innovation will lead to a creation of many more jobs' (Ford 2015). This opening of the bestselling book 'The rise of the Robots' summarizes the situation pretty well. We hope—but we don't know. Among the various studies and analyses currently available, it is unclear what the increased usage of technology means for the job market.

Digitalization is seen as a key influencer on future work concepts over the next decades. We can assume, that specific tasks performed by humans today will most likely disappear, if (ro-)bots can do the job as well. There will be new jobs emerging, as always, like Data Analyst and Programmer, but it is unclear if the number of new jobs will be higher or lower than the ones lost.

OECD estimates, that across all member states, approx. 10% of all jobs are automatable, another 15–35% of jobs will see significant changes in tasks (OECD 2016). Other studies operate with different numbers, but overall there seems to be some consensus that between 10 and 20% of all current jobs might be at risk, while another 20–30% are highly affected by digitalization (Dettmer et al. 2016). This means that up to 50% of the jobs currently performed today are highly affected and will see significant change in the coming years.

Changes in workplaces are not new. Some 200 years ago, 70% of Americans worked on farms, today it is less than 1% of the workforce (Schultz 2016). When machines took over the farm work, farm workers took care of Maintenance and Management. But today, we face a different scenario: This time, not only physical jobs are replaced by intellectual ones, but machines carry out more and more intellectually challenging tasks (Brynjolfsson and McAfee 2014).

Contrary to the past, it will not necessarily be the lowest-paid, lowest-qualified jobs that will disappear. There are different criteria at play now. A large number of more or less serious 'check-lists' are available to determine which jobs are at risk (see as an example Meyer 2016): 'Does your job rely on existing knowledge and existing rules? Do you perform repetitive tasks? Are there many people like you doing exactly the same job? Are you manually transferring data? Can your performance be acquired outside of the company?'—those are typical questions to check whether a specific job is at risk.

Who will take those Jobs? Automated systems and industrial robots are already common in Manufacturing settings and will add more and more intelligence over time. In Communication jobs, we have seen the rise of Chatbots (computer programs developed to simulate intelligent conversations with human users via auditory or textual methods) and Social Bots (a sort of Chatbot for Social Media to automatically generate messages and tweets). Especially Social Bots are capable of advocating specific content and ideas and can act as followers or pretend to be humans in Social Media. This includes the risk of spreading "fake news" (systematically planned and executed disinformation, i.e. for political campaigns).

4   Trust as a Key New Paradigm in a Sustainable Digital World

Trust as the Overarching Challenge

Most of the positive effects of Digitalization are currently still promises. Cleaner energy, higher productivity, shared economy, less resource-usage, and so on. Promises, people have to simply trust.

The previous chapter looked at three different areas of Technology Development and Digitalization—Data, Algorithms, Bots—that all are somehow interlinked and present both opportunities (mainly for Business) but also huge challenges for the way we want to live and work. Thus, they can be looked at as key influence parameters for concepts of Social Sustainability.

Focusing on the growing importance of Data and Data Management, a general fear for citizens lies in the perceived lack of privacy and intransparency of their data. It remains unclear what data are given (Splendid Research 2016), who is using the data and for what purpose, what is the value of these data and if customer even understand what data could be relevant for the store or service provider. It also remains unclear who controls the algorithms and what jobs are at risk. There is a significant lack of trust to companies, that is currently compensated with discounts and convenience, additional services and job enlightments.

While unclear usage models of personal data mainly concern individual users, the risks of any uncontrolled usage of algorithms in Social Networks is rather a general societal issue. Today, it seems very convenient to always get the famous '...customers who bought A also bought B...' message, it deprives users from surprises, from new stimulus and potentially new and positive experience. More importantly, it might exclude them from reality, i.e. political discussions that remain unseen, as voters believe opinions within their bubble. The risk of giving more and more power to algorithms who then decide as the programmers told them, but often with unintended consequences, is certainly rising for all societies worldwide.

Regarding the usage of Social Bots, we have seen significant influencing of the US Presidential Election 2016, where about 20% of tweets used during the campaigns came from machines (about 4 m tweets, hiding behind 400,000 fake identities) (Collett 2016) and we have seen attempted murder in the U.S., based on fake news generated by Bots (Kang 2016).

The impact of 'bots rising' for the labor market is unclear as of today. Most predictions estimate a disappearance of approx. 10–20% of today's jobs overall, but this could be up to 70% in some sectors, while others are barely affected. It also becomes clear that even today's high-paid jobs (like basic tasks of lawyers or doctors) can be threatened by bots. This will leave a significant level of uncertainty, both for the job market as well as for the individuals, and potentially a new level of distrust towards your employer. The impact of these three high-levels influencing factors of Digitalization can be visualized as in Fig. 4.

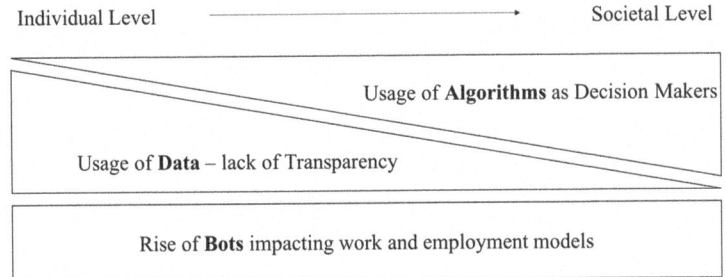

Fig. 4 Impact direction of key digitalization vectors

Need of Trust

With more and more data usage, people are increasingly afraid that privacy and security will disappear. More and more citizens seem convinced, that with increased importance of algorithms, diversity and broader knowledge will go away. And with the entrance of (ro-)bots, jobs and employability will vanish. Trust in the relevant institutions can change this.

Data, Algorithms and Bots present key components of a Digital Society, but the impacts need to be aligned with societal expectations. Trust is needed at individual level to participate in the data economy and understanding personal and professional opportunities and threats. It is needed in believing news and updates on Social Media that they really come from trustworthy sources. And trust is needed at the employment level to believe in employers and in own (maybe new) capabilities not to be replaced by bots soon.

Level of Trust

Over the years, with a small exception in 2015, we have seen a modest increase of trust towards business by consumers around the globe (Edelman 2016). However, there is one important aspect to pay attention to: The Informed Public (university graduates who follow the media and have incomes in the top 25%) is significantly more trusting institutions (Business, Governments, NGO's and Media) than the general population. According to Edelman, people who understand the changes and are capable or willing to adapt, are more likely to trust the changes that business and technology initiate. The more one understands the concepts of the 'New World' the more likely he is to trust the key actors.

It is also remarkable, that many people who say they do not trust businesses, have actually little ideas who *business* is and who are the people leading them. Name a CEO? 53% of people in the US could not name one, 68% of UK residents failed and 80% of German respondents did not come up with one single name. At a global level, only Mark Zuckerberg and Bill Gates received significant mentions (Edelman 2016).

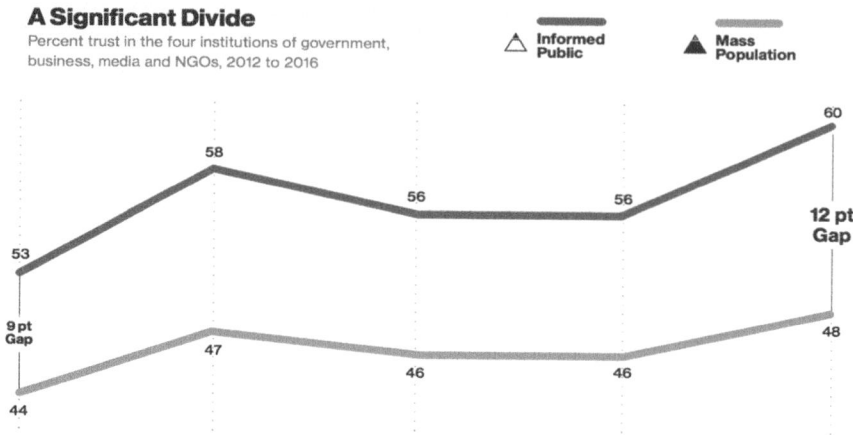

Fig. 5 Trust in Institutions among the Informed Public and the Mass Population (Edelman 2016)

Industry	2012	2013	2014	2015	2016	5 yr.Trend
Technology	76	73	75	73	74%	▼ 2
Food & Beverage	63	63	64	63	64%	▲ 1
Consumer Packaged Goods	57	60	61	60	61%	▲ 4
Telecommunications	58	60	61	59	60%	▲ 2
Automotive	62	65	69	66	60%	▼ 2
Energy	53	57	57	56	58%	▲ 5
Pharmaceutical	54	54	55	54	53%	▼ 1
Financial Service	43	47	48	48	51%	▲ 8

Fig. 6 Trust levels across Industry sectors 2012–2016 (Edelman 2016)

In understanding the widening trust gap, the information process also needs to be looked at—and Social Media plays a significant role: Today, general public is relying less on newspapers and traditional magazines (as the informed elite does), but choose self-affirming online communities as the most credible source of information. People active in social networks mention *friends and family* (undoubtedly with a similar value system) and *Search Engines*—as the predominant information source (Edelman 2016) (Fig. 5).

Looking at the trust levels by industry, we see that for the last years, trust in Technology firms was higher than for all the other sectors, though a little declining over the years (Edelman 2016) (see Fig. 6). This is especially surprising, as it seems that the industry driving change more than any other sector is at the same time the most trusted.

This, however, might not be the case in the future. Even the informed public is increasingly skeptical that the pace of innovation is at the right speed. Only 1 in 5 said it's right, but more that 50% of global respondents consider that innovations

TRUST IN BUSINESS INNOVATION:
PACE IS TOO FAST BY A 2-TO-1 MARGIN
THE PACE OF DEVELOPMENT AND CHANGE IN BUSINESS AND INDUSTRY TODAY IS...

Fig. 7 Trust in Business Innovation among the Informed Public (Edelman 2015)

INNOVATIONS PERCEIVED AS DRIVEN BY TECHNOLOGY, BUSINESS TARGETS AND GREED/MONEY

DRIVERS OF CHANGE IN BUSINESS AND INDUSTRY TODAY ARE PERCEIVED TO BE:

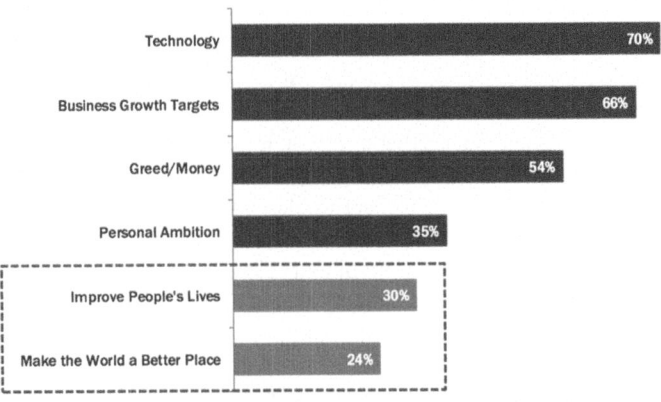

Fig. 8 Perception of importance of Innovation drivers (Edelman 2015)

come out too fast (see Fig. 7). This might be an indicator of potentially declining trust in the future.

Another aspect to worry about future trust to Business stems from the survey results that innovation is seen less and less being motivated to improve people's lives and make this world a better place (Edelman 2015).

Overall, we currently see a rather high level of trust into businesses, mainly in the IT Industry. There are two things to watch out for, though: The general public is trusting institutions much less than the informed public, indicating that more and more people risk of being 'left behind'. Overall, there are risks of declining trust to Business, as the motives for innovation center less and less around people and societies (Fig. 8).

Fig. 9 From trusting institutions to trusting individuals (Botsman 2016)

A New Model of Trust

Rachel Botsman, known for her research in the area of collaborative consumption (Botsman and Rogers 2010), where trust is also a dominant prerequisite, recently presented the concept of the 'Trust Stack', that is very helpful in understanding how trust can be improved at three different (building on each other) trust levels. The sharing economy is largely based on peer-to-peer marketplaces that depend on the social glue of trust between strangers. We are currently at the start of the shift from trusting people more than corporations or government (Fig. 9).

This new era of trust needs a measure, namely 'reputation capital' (Botsman 2016), which can be understood as the 'the sum value of your online and offline behaviors across communities and marketplaces.' It will transform how we think about wealth, markets, power and personal identity in the twenty-first Century—and it will be the key basis for societal trust in a digital world. Conventions of how trust is built, managed, lost and repaired—in brands, leaders, and entire systems are being turned upside down. Technology is creating new mechanisms that are enabling us to trust unknown people, companies and idea (Botsman 2016).

In order to build this reputation capital, a new trust framework is emerging in the collaborative economy, the 'Trust Stack'. In the first layer of the Trust Stack, people have to trust that a new idea is safe and worth trying. The next layer is trusting the platform, system or company facilitating the exchange. The third layer is all about trusting the other user while interacting with each other (see Fig. 10).

Fig. 10 The trust stack (Botsman 2016), own illustration

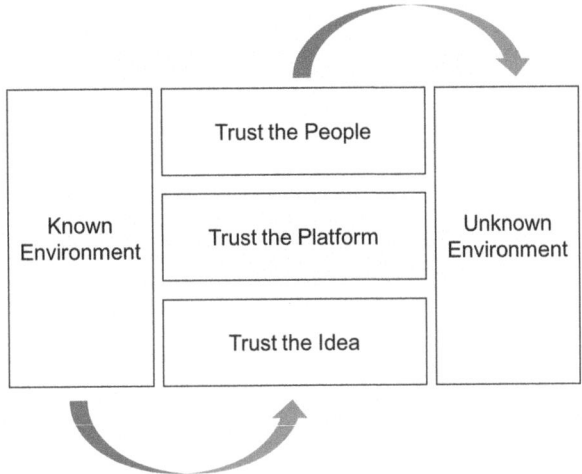

Over time, people open up to changing their behavior the more they 'live' in these trust structures, and then eventually regulations and policies adapt to ultimately change a system that is sustainable for society in a digital world.

5 Summary

This contribution took a look at three of the most relevant developments in technology, their impact on Digitalization and in the long run on how we want to live and work—Social Sustainability. We have only seen the beginning of it yet and the future is all but clear. Innovations happen at an ever increasing speed and new technology will continue to enhance our lives. At the same time, and not downplaying all positive outcomes of digitalization, we need a closer look and more focus on what the relevance and impact for society will be. Because it will affect how we act as a community, what values we pass on to the next generations and how sustainable our society is as a whole, in ways we want it to be. If we are not careful, digitalization might have impacts on how humans live together that can't be easily redone. We need an open discussion on the consequences of digitalization and we need transparency. As Digitalization continues, it mainly requires trust as a new glue. Not only for the informed Elite, but for all people. We all need to trust the ideas, trust the platforms, and trust the people behind. Over time, people are only likely to change their behavior if they 'see' these trust structures, and then accept changes in a system that is really sustainable for society in a digital world (Fig. 11).

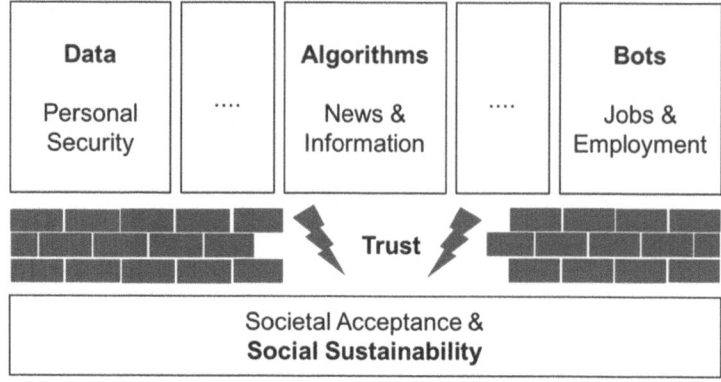

Fig. 11 Relevance of trust to move to sustainability in a digital world

Richard Edelman (2017) summarized the new challenge well: 'We have moved beyond the point of trust being simply a key factor in product purchase or selection of employment opportunity; it is now the deciding factor in whether a society can function'.

References

Botsman, R (2016) Currency of trust. http://rachelbotsman.com/thinking/. Accessed 22 Nov 2016
Botsman R, Rogers R (2010) What's mine is yours. The rise of collaborative consumption. HarperCollins, New York
Brynjolfsson E, McAfee A (2014) The second machine age: work, progress, and prosperity in a time of brilliant technologies. W. W Norton, New York
Clauß U (2013) Wie Gesichtserkennung den Einzelhandel verändert
Collett M (2016) A fifth of all US election tweets have come from bots. http://www.abc.net.au/news/2016-11-08/a-fifth-of-us-election-tweets-came-from-bots/8004726. Accessed 27 Nov 2016
Darwin C (1859) On the origin of species. John Murray, London
De Croo A (2015) Why digital is key to sustainable growth. https://www.weforum.org/agenda/2015/03/why-digital-is-key-to-sustainable-growth/. Accessed 12 Oct 2016
Dettmer M, Hesse M, Jung A, Müller MU, Schulz T (2016) Mensch gegen Maschine. Der Angriff der Roboter gefährdet die Existenz der Mittelschicht. Der Spiegel, issue 36/2016, pp 10–18
Domscheit-Berg D (2016) Die zwei Seiten der Digitalisierung. Horizont 37/2016, Special Edition, pp 10–11
Edelman (ed) (2015) Edelman trust barometer 2015. Edelman Trust Barometer, New York
Edelman (ed) (2016) Edelman trust barometer 2016. Edelman Trust Barometer, New York
Edelman (ed) (2017) Edelman trust barometer 2017. Edelman Trust Barometer, New York
Ford M (2015) The rise of the robots. Technology and the threat of mass unemployment. OneWorld, New York
Frey C (2016) Revealed: how facial recognition has invaded shops—and your privacy. https://www.theguardian.com/cities/2016/mar/03/revealed-facial-recognition-software-infiltrating-cities-saks-toronto. Accessed 21 May 2016

GeSI (2015) #SMARTer2030. ICT solutions for 21st century challenges. Publication of the Global e-Sustainability Initiative (GeSI), Brussels

Kang C (2016) Fake news onslaught targets pizzeria as nest of child-trafficking. NY Times. http://www.nytimes.com/2016/11/21/technology/fact-check-this-pizzeria-is-not-a-child-trafficking-site.html. Accessed 01 Dec 2016

Lindstrom M (2016) Small data: the tiny clues that uncover huge trends. St. Martin's Press, New York

Meyer J-U (2016) Hilfe, ich werde wegdigitalisiert. http://www.manager-magazin.de/unternehmen/karriere/so-erkennen-sie-ob-die-digitalisierung-auch-ihren-job-bedroht-a-1113678.html. Accessed 02 Nov 2016

Müller von Blumencron M (2016) Wieviel Privatsphäre sind wir bereit zu opfern? Horizont 37/2016, Special Edition, pp 10–11

OECD (2016) Automation and independent work in a digital economy, policy brief on the future of work. OECD, Paris

Osburg T (2013) Social innovation to drive corporate responsibility. In: Osburg T, Schmidpeter R (eds) Social innovation—solutions for a sustainable future. Springer, Heidelberg

Osburg T (2017) Corporate social innovation und Unternehmensstrategie. In: Wunder T (ed) CSR und Strategisches Management. Springer, Wiesbaden

Pariser E (2011a) Beware online "filter bubbles", TED-talk. https://www.youtube.com/watch?v=B8ofWFx525s. Accessed 02 June 2016

Pariser E (2011b) The filter bubble. What the internet is hiding from you. Penguin Press, New York

Rosenbach M (2016) Mit dem Ich bezahlen. Der Spiegel, issue 11/2016, pp 82–83

Sächsische Hans-Carl-von-Carlowitz-Gesellschaft (eds) (2015) Zur DNA der Nachhaltigkeit: Carlowitz weiterdenken. München

Schultz S (2016) Arbeitsmarkt der Zukunft: Die Jobfresser kommen. Der Spiegel, issue 32/2016

Splendid Research (eds) (2016) Studie: Bonus- und Vorteilsprogramme 2016. Hamburg

Steinbrecher M, Schumann R (2015) Update. Warum die Datenrevolution uns alle betrifft. Frankfurt am Main

Tardieu H (2014) Sustainability and digital revolution: a winning combination. https://ascent.atos.net/sustainability-digital-revolution-winning-combination/. Accessed 12 Oct 2016

TNS Infratest (2016) Digitale Begriffe für Bundesbürger noch immer Neuland. http://www.tns-infratest.com/presse/presseinformation.asp?prID=3474. Accessed 30 Sept 2016

United Nations Global Compact (2016) Social sustainability. https://www.unglobalcompact.org/what-is-gc/our-work/social. Accessed 17 October 2016

Wadhawan J (2016) Kein Wert für alle. absatzwirtschaft 04/2016, pp 26–37

Weingarten S (2015) Der neue Souverän. Wie Facebook, Twitter und Co. den Staat überflüssig machen könnten. Der Spiegel, issue 27/2015, pp 66–67

World Economic Forum (WEF) (2011) Personal data: the emergence of a new asset class. World Economic Forum, Cologny

The Risk Averse Society: A Risk for Innovation?

Stefan Schepers

1 Introduction

The increasing differences between industry and EU institutions and governments about risk assessment and risk management, the introduction of a precautionary principle in the EU Treaties and the move towards a hazard based approach are based on deep rooted cultural changes in Western society and, in Europe, on particular political mechanisms. This general trend is behind most of the European consumer protection, health and environment legislation, while the emergence of hazard based regulation is most prevalent in the agro-food and chemicals sectors, but it is creeping into other sectors too and nothing guarantees that one day it will not affect the digital and other new technology sectors. The ever more politicized use of the precautionary principle is a barrier to economic and social innovation.

These developments are the result of a growing widespread doubt about the advances of science and technology which are seen to produce news risks which differ in character from risks of an earlier industrial age. These new risks, manufactured by industries in various sectors, are seen to potentially affect everyone and they are creating therefore a high degree of social uncertainty, which influences politics and market conditions. They are not necessarily risks to physical health or to the environment, they include also new risks to private rights produced by the business models in the digital sector or by the collateral effects of some ICT technologies and their use by companies or governments. A number of other specific cultural circumstances in post-industrial western societies make people particularly apprehensive for these new manufactured risks.

S. Schepers (✉)
EPPA, Brussels, Belgium

Henley Business School, Reading, UK
e-mail: stefan.schepers@eppa.com

T. Osburg, C. Lohrmann (eds.), *Sustainability in a Digital World*, CSR,
Sustainability, Ethics & Governance, DOI 10.1007/978-3-319-54603-2_2

These weaken the credibility of science, industry and government. Public suspicion about governments' ability to deal with danger and risk is reinforced by inefficient risk communication strategies, as governments still tend to rely on outdated models of risk communication, which "tend to define the public as an essentially naïve audience" (Botterill and Mazur 2004). Risks are inherent to economic and technology developments, but they are difficult to deal with and require a careful, long term strategy in order to at least reduce the potential political and social antagonism, which nearly always translates in electoral agendas and in market regulations. In particular the EU institutional system seem to be picking up the new cultural paradigm, for reasons of its own.

Sociological analysis has shown that scientific argumentation, the cradle of most business communication about risk, does not suffice, because the new manufactured risk concern is itself a construction of scientists. Industries must learn to take these cultural attitudes seriously and to develop a new approach towards them, first in communication but also in their strategy, dependant on the specifics of each industrial sector.

Regardless how great technological opportunities, if people do not want it, for rational or irrational reasons, resistance in the commercial and in political systems will be such that formidable obstacles will emerge, up to banning specific use. Sometimes this will be justified for ethical reasons or for the protection of civic and human rights, or of social protection systems which in particular in Europe are extensively developed; and they are a key element of political and social stability and of leveling of the boom and bust cycles in the consumer driven economy.

But when it is based on improper risks assessment and management, caused not by the technology itself but by socially insensitive and politically unwise handling by corporations, then the cost benefit balance becomes distorted. This risk is growing all the time because of the rapid advances of science and the cognitive gaps which this causes with decision makers and the general public alike (Dror 2015).

The Challenge

Industry has been struggling since at least a couple of decades with a new political and regulatory approach towards risk and the move towards a hazard based approach. It has been mostly on the defensive, arguing from a traditional scientific perspective which is no longer sufficient because of the cultural shifts which occurred meanwhile.

Civic society organizations, and an increasing number of political decision makers in all political parties, are moving towards a hazard based approach and possibility risk assessment instead of a probable one, putting additional pressures on short term commercial interests and long term business models. It increases also massively the regulatory burden on industry which has to carry out ever more scientific studies and move through complex time and manpower consuming procedures, in particular in a multi-layered governance system as the EU, in order to get market access. It may upset business models themselves, simply because it is more difficult to argue about a possible risk, a hazard, to say, privacy, than about a probable one; it reverses almost the burden of proof.

In order to develop political compromises and fundamentally better regulation, not just cosmetically, between industry and political decision makers, and to be able to improve its societal dialogue, it is important to understand the deep rooted causes of this cultural shift (Beck 1986; Giddens 1990).

At the very least, this apparently unstoppable social and political development requires industry to fundamentally change its approach to risk communication, though it will require over time reconsideration of R&D and of commercial strategy. Like some corporations are doing already, non-market, 'soft' issues will have to move centre stage in strategy development and implementation (Polman 2014). Instead of arguing its case from a research risk approach, it needs to fully include a social research risk perspective. Failing these inputs, dialogue efforts will have little to no effect with the vast majority of people, including decision makers and all kinds of vested interests threatened by innovation.

2 From Industrial Risk to Risk Averse Society

Risk in Historic Perspective

The contemporary concept of risk is a product of modern times, it did not exist before.

From times immemorial until the beginning of modernity, people knew only dangers (hazards) as natural phenomena, such as hunger, illness, earth quakes, floods etc. They were attributed to metaphysical powers (gods and demons), which existed outside humanity but influenced it, favourably or not. They were 'natural' events, an inevitable part of the seasonal cycles and of life, steered by these divinities. In any case the thought that these adversities could be caused by humans did not occur at all; in fact, they seldom were. Religions gave people understanding and meaning and helped their acquiescence in the face of dangers and catastrophes, not least by promising a better after-life and reunion with loved ones. The absence of a risk concept excluded thus the idea of human responsibility, people underwent the dangers as an inevitable part of human existence.

This started to change with the sixteenth and seventeenth century discovery voyages and the uncertainty of safe return and of a commercial result. Risk enters European culture together with the maritime adventures: finding new land and re-paying the capital invested became dependent on natural and human causes; not only the tides and winds determined the outcome, but also the ship's construction, new technology, and the ship's management. Moreover, the Renaissance gave new impetus to Greek philosophy and made people aware that they could take their own destiny in their hands, a view which merged well with Christianity, but was absent in this sense in other religions.

Risks, as we consider them today, emerge fully with the beginning of industrialization, during the nineteenth century, which brought dangers which had no longer a natural cause but which were manufactured, human made dangers, byproducts of industrial process and of technology. A driver losing control of a car is not the same as a horse which suddenly rears. The modern risk concept is based on science and technology and on new ideas in mathematics concerning probability. Dangers moved now from the realm of divinities fully into the realm of humans and it became a part, not of nature, but of science, technology and industry. An earth tremor is not the same if it is caused by shale gas exploitation or by the slow movements of the tectonic plates of the planet.

However, it is important to note that a manufactured danger is not purely objective and value free; on the contrary to become a risk in the modern sense of the word is needs to be constructed as a social fact. (Beck 1992) This can happen when there is a wide gap between the technical knowledge of the scientist and the subjective understanding of ordinary people, which may over- or underestimate a risk. The acceptance of such a risk construction depends very much on the cultural and social context and on gender; women are usually more risk averse, so are people who feel uncertain or threatened by social, cultural or economic shifts.

Often risks enters popular culture as a result of a major accident or of a sudden unpleasant discovery, such as the indiscriminate NSA spying on people which exposed to many the risks of the digital age, but in itself this is not enough : there needs to be a science based construction which counters the industrial risk analysis. But even then the new manufactured risk needs to correspond to several characteristics for the risk to become a permanent and policy influencing feature in society. Therefore, just better or more communication with the general public will not help much, because the acceptance of a manufactured risk as a new hazard depends on a cultural, social and ethical context in society (Lupton 1999). In this respect, Europe diverges from other parts of the world.

From the late nineteenth until the mid twentieth century, the emergence of manufactured risks was accompanied by a whole system of institutions and regulations to manage them. For a long time, this system was able to provide sufficient assurance to the general public and to let industry and science operate more or less un-opposed.

Social science has shown that at that development stage of industrial risks, the consequences of risk for individuals were researched, calculated and two socially re-assuring responses were elaborated over time: prevention and compensation (Beck 1992). Both are a systemic solution by governments to deal with the social and economic uncertainty resulting from the hazards of technological and industrial developments. Both have the advantage to make political and legal conflict unnecessary and thus also to reduce social instability, a threat to governance and business alike. These insurance mechanisms, private insurance as well as public one (the welfare state social and health security system) are an essential compensation for the uncertainties created by modernity, they are a cornerstone of the social agreement between industry, government and society.

Some sociologists argue also that the present risk prevention approach is based on the same modernist utopia of being able to determine all aspects of life through science and technology, including thus the hazards created by the very same modernity of industrial and post-industrial society. Failing to prevent in an absolutist certain way leads then to doubts about the promises of science and technology themselves (Castel 1991). It increases again feelings of social anxiety, for example as a result of chemical or nuclear accidents or of food crises, which we thought to have overcome but which seem to come back (or have never been away). The same applies to new risks, such as those posed by digitalization to privacy.

More importantly, many social scientists seem to believe that in the end, the new risks are not constructed by lay people, but by experts themselves, either from competing economic sectors, or from civic society organizations (with an interest in funding and members) or from university (with an interest in public research funding), or from specialist but oligarchic international organizations (such as WHO), or for purposes of electoral engineering by political parties. They are then magnified in the media and start to have a life on their own. It is also acknowledged that this risk construction as a social phenomenon is part of a backlash against big corporations and part of the decline of trust in government (or in EU institutions).

It must be noted that big corporations have contributed themselves to the erosion of the trust in compensation mechanisms, by endless judicial procedures, and sometimes other actions, against groups or individual citizens and the avoidance of compensation payments.

However, in addition to the reasons given above, the new manufactured risks alone would perhaps not have led to the emergence of the so-called 'risk (averse) society', if they had not been accompanied by two deep rooted cultural changes: a new relationship to nature and individualization.

Industrialization was based on the premise that humans and science could dominate nature and put it to their advantage. But we live today in a post-nature time, we now are primarily concerned about what has happened to nature and about the consequences of our interventions (e.g. climate change). This fundamental change is also an element of the transition to the risk society (Giddens 1990).

In addition, our time is characterized by the so-called end of tradition, where people do no longer expect their lives to be per-determined, but to be able to make their own life according to their views and wishes. This was of course part of the expectation of Enlightenment philosophy. Its realization, in the Western world at least, has led to a high degree of individualization, which strengthens the anxiety about new, manufactured risks and their potential indiscriminate consequences.

Western modern humans (want to) feel individualistically in charge of their own life, but at the same time they feel isolated from others (decline of the family and other protective bonds provided by the state) and facing new dangers resulting from the technologies which they nevertheless favour and need to conduct their life at the high material standard of Western Societies. In Europe, the advance of secularism has also played a role because divinities have been eliminated from our explanations of danger and risk.

New Risks and Their Effects

In the second half of the twentieth century, this earlier social agreement starts to show cracks as a result of a new category of risks. These are influenced by different set of events in different parts of the western world and they lead to different responses. But fundamentally they are all part of the new risk society. In this era, risk is no longer seen as potential source of human progress and technological innovation, yet it rather entails negative connotation.

In Europe, the fear of nuclear catastrophes grew because of accidents in several places, the worst one in Chernobyl (1986), which caused radio-active fall-out and related illnesses (cancer) over a large area and for a long period. Then came several ecological catastrophes, such as the chemical explosions in Bophal (India) and Seveso (Italy), the oil tanker accidents in Alaska, Normandy and elsewhere. These are followed by several health and food scares (e.g. the HIV contamination of blood for transfusion, the BSE scandal, the dioxine crisis, the threat of pandemic diseases such as avian flu).

In the USA, it is claimed that the assassination of President Kennedy, and later the Vietnam war, have triggered a decline in the belief of progress, though it has not been as pronounced (yet) as in Europe, where these developments came on top of a growing belief of losing one's dominant role in the world. More recently, the threat of terrorism (often exaggerated for political reasons) must be added as a new source of social anxiety. The latest to join this (non-exhaustive) list is the digital sector and its real or presumed dangers to privacy and consequently to civic and even human rights, to the relationship of people with each other, and with the government. It are not always ordinary people which are worried, data ownership is an issue of huge commercial and political importance.

While in the USA social anxiety has been channeled very much into the revival of fundamentalist churches and their escapist offer from modernity (or certain aspects of it), an understandable reaction given the puritan origins and the equally strong belief in entrepreneurship, the response in more secular Europe has been to more focused on precautionary regulation by public authorities, also understandably given the traditionally greater interventionism of public governance in the economy. However, in the very recent case of new financial risks, one sees a convergence of approach.

As analysed by the social scientists mentioned, these new risks, of an ecological, health, lifestyle, economic or criminal nature, are fundamentally different in the general opinion from the early industrial risks of the nineteenth and first half of the twentieth century:

1. The distribution of risks has changed, they are no longer limited in time and in location and, very importantly, they affect indiscriminately all social classes. Some risks, such as the effects of climate change or of a nuclear catastrophe, are even global in nature. In particular the electorally important middle classes, whose economic conditions are made already more vulnerable by the effects of a

globalizing economy, believe that their living conditions generally, and those of their children, are more uncertain than before, because of the many new risks.

2. They are man-made risks, part and parcel of modern society, with all its accepted benefits, by the joint action of science and industry. As Beck, Giddens and others explain in depth, they result directly from the exploration by science and the transformation by industry of nature by human activity. But they are potentially worse than previous, natural disasters, they can even lead to a 'worst imaginable accident'. In more recent works, social scientists add to these nuclear, ecological and chemical new risk categories also genetic risks (all genetic engineering), robotics, and perhaps foremost artificial intelligence. They are producing new dangers and new uncertainties for which we do not yet have a coherent regulatory architecture nor in many cases a clear ethical consensus (Dror 2015).

   The risk of the digital sector should be added to these concerns, though they may cause no immediate ecological risks (except in the mining regions of rare earth minerals), they cause new, diffuse risks to people's life, such as unknown invasion of privacy, and next to opportunities also threats to European culture and the nature of its societies (Praet and du Puy 2016).

   The risk aversion about which these social scientists write is thus no longer limited to environment or health and to specific industries. They apply it also to the global financial services sector, in particular because it creates new welfare and employment risks, affecting both the working and the middle classes, thus adding to the 'culture of fear' (Furedi 2005). Also the security fears of recent are seen as contributing to this, not least because they are used by many governments to enlarge control mechanisms over citizens, without their knowledge and consent (and using digital technology), and by businesses in the security sector to enlarge their markets.

3. The scientific and legal responsibility of these new risks is very difficult to attribute. As a result, industry got away in most cases with impunity or very low damage payments, even when the suffering of people was real. People feel helpless against the legal battalions of a corporation; even an apology is often too much because of potential legal implications. But what corporations win in this way, they lose in another way: there is no free trust, no free social contract, there is always a price to pay. The insurance mechanisms of old thus no longer function. Corporate lawyers can do a lot of damage to their company social contract.

   Also governments have often tended, until recently, to minimize, sometimes selectively, the consequences of the new risks. The new risks undermine the basic concepts on which the modern state and its economy are build: its legal system, the political institutions, and the key economic institutions (companies of many sectors).

   There is little chance of redress, therefore it is often more likely to spread fear and risk aversion, also because of potential abuse of the new technologies by business and governments alike. There are equally warnings that the sovereignty of states will be further weakened, if not undermined, by the combined effects of ICT and global financial markets. Like in other industry sectors coming under social scrutiny before, the dominant culture itself of the IT sector, hyper-individualistic, anti-social and anti-government, may work against the solidity of its social contract (Turner 2014).

Ultimately, this is bound to lead to government and judicial interventions, which in Europe clearly have started. Corporations usually go to three stages then, from denial to counter-propaganda and lobbying to acceptance to deal with the collateral effects of its activity by collaborating, the more clairvoyant ones at least, with governments and EU Commission to bring about a balanced regulatory architecture.

4. The perception of these new risks is different from previous ones. Risks of old were accepted by non-experts because they believed that government and industry were able to control them; this belief has changed into doubt (and this is itself one of the watersheds between modernity and post-modernity). Consequently, people are more open for doom preaching organizations, focusing on worst cases and hazard. The often existing divisions among scientists add to this uncertainty.

It are experts themselves which often start to harbor doubts or ethical concerns about the consequences of a technology, more often though about the use being made by a corporation or a government. The revelations by people like Edward Snowden and others will have long term consequences for the digital sector. Indeed, it may have comparable effects on the digital sector as the work of Rachel Carson had on the chemical sector (Carson 1962). Now like then, the social warning signs, and some scientific facts, were ignored and a heavy regulatory burden descended on the chemical industry. In both cases, their moral courage was recognized by many people, contributing to the weakening of trust in the industry.

The IT sector is one which still lives in the phase of hubristic denial and antagonistic lobbying; this relative absence of interest in the social contract, in the collateral side-effects of its business activities, is bound to increase risk aversion about it. This is strange, because each industry sector, given sufficient inclusive strategic thinking, usually has itself the scientific and technological remedies for the concerns raised, as shown for example by the chemical or the vehicle manufacturing sector where the competitiveness of some corporations, those with an extrovert culture, has benefited from doing so. But foresight and alignment with stakeholder concerns and interests have too seldom played a significant role in corporate strategy driven by quarterly results (Schepers 2011).

5. The communication about the new risks by the media is such that the cause-effect relationship is often not understandable to the general public. This is a general problem, resulting from the packaging of news into small, rapid items (the CNN model), but it comes on top of the other characteristics of the new risks.

The result of these developments of the last 40–50 years is a serious weakening of the existing social contract. The new category of risks therefore does no longer benefit from social acquiescence. Whereas there was a previous belief that risks could be brought under control, this is breaking down now and is being replaced by scenarios of doom with various degrees of plausibility. Governments and industry have lost credibility. What is worse, so has science itself: greater knowledge has led to greater uncertainty (Giddens et al. 1994).

In order to fully apprehend why this could happen in just about one generation, one has to take into account that modern western society has become more individualistic. In highly developed welfare states, people have been receiving more rights and entitlements, but only as individuals; only recently has there been a start with corrective mechanisms, for example in the health sector by the introduction of (small) payment to families caring for the sick or elderly, which reliefs the financial burden on public health systems and brings psychologic benefit to the person concerned. The comforts of old of the family or the social group to which they belonged have been seriously weakened. But the new opportunities in an ever more complex society have created now 'freedom risks' (Beck 1995), i.e. one is supposed to take decisions (e.g. to eat or not to eat GMO food, to support of nor nuclear energy, to give personal data away on the internet) of which one does not and cannot know the future consequences (risks). Traditional comforts for the individual are weakened or have disappeared.

Ultimately, the new risks are such that they can constitute a risk to society itself, as conceived since eighteenth century Enlightenment in the western world. They are no option, they are seen as an unavoidable by-product of the progress of science and industry, which now has generally weakened, or even undermined in some cases, itself. And this despite the fact that science and technology are often able to provide the solutions themselves, but this is often done belatedly because of existing business models and their single-minded focus on short term shareholder value.

The EU's Precautionary Principle
The new cultural attitude has led to a new approach to risk management in the EU in particular, based more and more on hazard and on possibility risk assessment.

Sociologists believe that the very existence of the precautionary principle leads, in the present cultural climate, to speculative thinking about worst-case scenarios and undermines the traditional probabilistic risk assessment. Societies have always been apprehensive about the unknown, but presently there is a belief that the unknown, unpredicted or unpredictable threats are even more dangerous than the known ones (Bauman 2006). Moreover, we have a cognitive gap about likely future threats, in particular from new and unknown technologies. The institutionalization of the precautionary principle in the Amsterdam Treaty (1999) thus leads to social and policy approaches based on the worst hypothesis, a consequence of people's difficulties of understanding and thus of interpreting the present world.

However, given that we are dealing with manufactured risks, the source of danger is no longer the lack of, but precisely knowledge itself (Giddens 1998; Luhmann 1993). Society in post-modernity has moved towards discomfort and uncertainty about scientific and technological progress and its associated risks. Whereas in the past people believed that one was capable of calculating risks, today we stress the inability to do so. It must be emphasized that this attitude is based on political developments in the twentieth century and the many failures to deliver on promises, it is strengthened by the various ecological and health crisis or scares, all of which undermine the trust of people in social institutions and lead to uncertainty.

More recently the authority of knowledge has been further undermined for political purposes, notably by the US and UK governments during their attempts to justify the Iraq war. They introduced to the public the concept of the 'unknown unknowns' (in addition to the traditional known knowns and known unknowns). It is used nowadays to justify anti-terrorist measures affecting the general public, but this is just another new risk, as the financial meltdown risk, which adds to the climate of uncertainty (Furedi 2009).

In addition to the already noted consequences of globalization, it strengthens a form of cultural pessimism in the western world It leads to the 'err on the side of caution' policy approach, as the EU Commission advocated, and to a possibility approach to risk. Probabilistic risk assessment is consistently devalued by what Furedi calls 'fear entrepreneurs' (to be found in civic society organisations, media, politics, and in businesses with benefiting from it, such as the growing security sector, or, in the case of food, the bio-agricultural sector). Interestingly, he quotes an American study which shows similarity of language between the US speak on terrorism and EU talk about ecology (Table 1)

Table 1 Overview of the changing perception of risk elements in different types of society by Beck (1995)

	Pre-industrial society	Industrial society	Risk society
Examples	Natural catastrophes, illnesses	Occupational risks, traffic accidents, etc.	Man-made catastrophes
Dependence on human decisions	No, the will of god	Yes, industrial developments	Yes, industrial developments
Chance to avoid	No, result of fate	Yes, e.g. safety belt, healthy living, etc.	No, collective decision with no individual chance to avoid
Who is affected	Whole populations, countries	More limited in time and scope, social limits (e.g. workers)	No social limits (everyone), uncertain effects intime and scope
Calculation, cause-result, insurance	Accepted insecurity	Calculable insecurity, compensation	Not calculable, difficult precaution, worst case scenario possible

3 Consequences on Governance Institutions

Following Foucault's approach on government and social disciplining, the specific role and interests of the EU institutions in dealing with the risk society show why Europe is moving from risk assessment and management to a hazard approach which creates potential barriers to the research and innovation value chain.

Undermining of Social Institutions
Nearly all sociologists analyse the consequences of these scientific and industrial developments producing the new types of risk on the social institutions (government, judicial system, scientific bodies) which make up the modern state. They have to identify, evaluate, communicate and regulate the new risks.

Until the mid twentieth century, the traditional risks manufactured by industry could be adequately managed through a system of causality, responsibility and insurance (Beck: if a fire breaks out, the fire brigade comes, and the insurance inspector). This system became upset when it became nearly impossible to ascribe the effects of new risks, to calculate the damage and to control the causes and the consequences. By their very nature, the new risks limit their own risk management because of their potential magnitude and their higher degree of uncertainty about effects in time and scope. Moreover, the old re-assurance mechanisms introduced by the state do no longer work as effectively, due to the nature of new risks.

In fact, the present methods of risk assessment are no longer adequate to deal with the new risks: there is still the traditional legal requirement to prove a causal effect and to prove damage by the victims, who have mostly not the means to do so in face of a diffuse ecological or health or privacy threat. Who can one legally prove that one has become ill due to traffic emissions, even if it is medically possible or even probable? Industry lawyers can easily dilute responsibility over many actors, with the result that many risks, even known ones, become un-attributable.

In addition to the uncertainty, this lack of redress creates a feeling of helplessness among citizens (compare the immediate and massive help to the earth quake victims with the lack of appropriate compensation for the victims of Bophal, or the Erica, or Chernobyl, even decennia after the accidents). The cultural impact of these new risks is thus different from the traditional risks: while the danger is potentially larger, and not defined in time and scope and class, people are on top of this in practice uninsured against it. This mismatch cannot but affect the credibility of public institutions which are normally responsible for this.

Influential political scientists, such as the late Lord Dahrendorf (a former EU Commissioner), have warned long ago that the (manufactured) climate of fear and uncertainty and the way in which governments deal with it can slowly undermine liberal democracy and that it can lead western countries towards more authoritarian forms of government. This danger has clearly increased by the response to terrorist threats in many countries, witness the state of emergency declared in some countries; it is the democratic equivalent of banning a chemical substance because of very rare risks of toxicological effects. Undermining civic rights will in turn undermine sooner or later market freedoms for industry too, because companies are legally created social institutions, and thus subject to and dependent on constitutional and civic rights, comparable to individual citizens. In fact, it is already happening.

Impact on Economic and Social Context

The risks produced by new science and technology undermine thus the social trust in science and in the institutions for public governance.

In traditional industrial society, the key social issue has been for a long time the just (re-) distribution of welfare, and the management of known (understood and more or less accepted) risks. In the post-industrial society, which by definition is a highly developed and wealthy one, the social logic changes from distribution of social 'goods' to avoidance of social 'bads'. This is a fundamental undermining of the principle of modernization of the last two centuries, with vast consequences for the economy, for scientific progress and for the system and objectives of public governance. However, many politicians have not yet changed mentally towards this new cultural paradigm, thus further weakening the political and, indirectly, industrial credibility (because of presumed collusion) and strengthening the role of civic society organizations opposing the present logic of producing new risks.

We are thus in an intermediate phase between the old left-right political opposition of industrial society and a new one which will have to focus on the new real or perceived challenges. In it in this gap that many politicians (in the Commission and Parliament) are trying to find the new credibility and acceptability of the EU.

Its effects are not limited to the above, they also change the industrial logic itself between those who benefit from the production of these 'social bads' (e.g. the nuclear industry) and others who offer competing products (e.g. the wind or solar energy industry). Economic interests emerge which benefit from maintaining the fear about new risks because it helps them to shift policies and regulations and consumption patterns in their favour (no clearer example today then the collusion between certain governments and the security industry in maintaining a high level of fear about terrorism).

There are also industrial sectors which can win twice in the risk society, by developing products destined to deal with the results of other products. In general, new risks resulting from production (e.g. ecological effects) are the ones which are most difficult to attribute and thus to manage as traditional risks have been. Product risks are something different, because they can rapidly destroy value and are therefore mostly better regulated, indeed companies themselves seek to avoid them.

Over time, the main driving force in the market economy may start changing from the provision of goods to the avoidance of risks. Industrial societies were primarily concerned with re-distribution issues, post-industrial societies are focusing on the avoidance of risks (new ones in the first place). In particular German sociologists have emphasized that the risk society strengthens the class society, because the lower classes have less information and less possibilities to avoid them;

it is indeed well known that ecological and health concerns are primarily a middle class phenomenon, though it is spreading to the working classes now too, in particular in Europe through the trade unions.

The EU and the Risk Society

While the rest of the world is moving towards a *risk management* culture, Europe is increasingly moving towards a *risk avoidance* concept, and the EU is often seen as contributing considerably to enhancing a climate of risk aversion in Europe and playing a major role in changing the quality and dynamics of European regulatory policies. When it comes to dealing with risk, Europe's Member States show different pictures and are geographically polarised: while countries such as Germany, the Netherlands and Denmark, for example, have a relatively stringent approach towards risk and risk management, countries such as the UK, France and Italy are less concerned and have less restrictive regulations in place (though France is moving now towards a northern European approach).

The original mandate of the EU was based on post-war economic reconstruction and on ensuring lasting peace between its members. It has been very successful on both accounts. The realization of the Single Market followed by the Economic and Monetary Union (and the Euro) was the last step in this process. The EU carried out the re-construction of economic sectors, with all associated tensions, thus freeing national governments from (part of) the difficulties in doing so; it allowed them to 'hide' behind a collective decision.

The end of the cold war has left the EU without a new overriding objective. Under German influence, the precautionary principle was written into the Amsterdam Treaty (1999) to be used in future environment and health policy. Its application was left open.

Ever since, the two institutions which have most problems to build or to maintain political credibility with the citizens, Commission and Parliament, have been tending to focus on the key concerns of the risk society and seeking to meet them. Thus the EU has become a leading political force for limiting the new risks, wherever its regulatory activity allowed it to do so. It is trying to find a new, or at least an additional, raison d'être in alleviating the new social fears by moving away from the traditional risk based towards a hazard based approach.

However, social institutions also have a role in disciplining and regulating population, and concepts of risk can be used to that purpose (Foucault 1991). In modern western neo-liberal societies, individual freedoms (including for business) are favored against government interventionism. Yet governments need to maintain a form of disciplinary power through various techniques in order to fulfill their functions. From this perspective, the new manufactured risks provide a useful tool for government's regulatory power exercise while still operating within the confines of neo-liberal thought. Recently some sociologists have explained the government responses to terrorism in these terms too. In addition to formal regulations, new manufactured risks can also be used for promotion of self-discipline (e.g. life style risks).

Here lies a second interest for Commission and Parliament to focus on health and environment. At a time when national governments are mostly resistant to transfer new powers to Brussels, and that the principle of subsidiarity has also become enshrined in the Treaties, focusing on the popular fear for the new risks has, given their characteristics, a disciplining effect on governments. They find themselves in a politically difficult position towards their own electorates if they are seen not to care as much as the Commission or Parliament. But national governments are not just losers: the focus on risks and the move towards hazard helps them to placate national industries (collective decision making by 'Brussels') while simultaneously placating the national risk minded opinion. So it helps in fact national political establishments too (Schepers 2016).

4 Conclusion

Sociological studies about risk show that the new approach towards 'risk equals hazard' is not a passing phenomenon and that it is not the result of some civic society campaign directed against a specific industry. All industry sectors are confronted with the consequences of deep rooted cultural changes in western society which have been emerging since the last quarter century and which will not go simply away with a new regulation or a large counter-campaign.

This new cultural paradigm, which like always is not uniformly spread in society or between countries, given that attitudes to manufactured risks vary with the cultural context, can in fact offer new opportunities for research based companies, but only if these new social attitudes are taken into account in corporate strategy and if they themselves are dealt with in a long term perspective.

It requires corporations to reach a clearer understanding of firm competitive advantage within new social and cultural paradigms. Thoughtfully articulated competitive strategy acts as the foundation for both the design of the organisation and the quality of its value delivery and for sustainable business models, asset value growth and profitability.

With ever fiercer competition attention has to focus on the intangible assets side to corporate strategy, in particular its societal and political context. Excellence in the development and the delivery of products and infrastructure, namely tangible assets, is increasingly taken for granted.

Enhancing company reputation and displaying accomplishment in the management of intangibles has now become a fundamental aspect of corporate strategy. Globalisation makes the political and democratic environment more uncertain and the regulatory environment more complex. The effects of the crisis increase again the role of governments in markets. This demands a fundamental rethink of strategy elaboration and implementation. The challenge for top management today is to align contrasting stakeholder agendas with the firm's commercial objectives. A critical component of this alignment is taking account of government policy objectives and its regulatory elaboration in such a way that the company turns such challenges to distinct competitive advantage.

Governments are challenged in developing new methodologies and tools of engaging with critical citizens, traditional and new media and business in order bring out alignment of views and interests and to realise the Common Good, a concept all too often overlooked in today's fragmented post-modern societies. The effects of policies leading successfully to economic growth over the past decades, research and technology developments, globalization, and the interactions among these, have led to more complexity in the society and in the economy than ever before. Therefore constant attention is needed whether a governance system is sufficiently adapted to the outcomes of its own actions and non-actions, to the co-evolving economic and social networks and their sometimes different phases resulting from different pressures in parts of a system such as the EU or the global market.

Public governance and corporate management innovation in line with new and continuously changing contextual conditions and new complexity are a permanent necessity, and a key part of countries' competitiveness and prosperity, as much as checks and balances within the public sector and between it and the private sector or inclusive policy making focused on the Common Good. Among other signals, the continuous growth of a risk averse society in the Western world, though with different emphasis and outcomes on both sides of the Atlantic, show that there is a long way to go before reaching an economically and socially new balance.

References

Bauman Z (2006) Liquid fear. Polity, Cambridge

Beck U (1986) Risikogesellschaft, auf den Weg in eine andere Moderne. Suhrkamp

Beck U (1992) Risikogesellschaft, auf dem Weg in eine andere Moderne. Sage, Thousand Oaks

Beck U (1995) Ecological enlightenment, politics of the risk society. Prometheus Books, Amherst

Botterill L, Mazur N (2004) Risk and risk perception. Report for the rural industries research and development corporation, Canberra

Carson R (1962) Silent spring. Houghton Mufflin, Boston

Castel R (1991) From dangerousness to risk. In: Burchell G, Gordon C, Miller P (eds) Studies in governmentality. Harvester/Wheatsheaf, London

Dror Y (2015) Priming political leaders for fateful choices. Eruditio e-J Worlds Acad Art Sci 1(6): 40–49

Foucault M (1991) Governamentality. In: Burchell G, Gordon C, Miller P (eds) Studies in governmentality. Harvester/Wheatsheaf, London

Furedi F (2005) Culture of fear. Continuum International Publishing Group, Harrisburg

Furedi F (2009) Precautionary culture and the rise of possibilistic risk assessment. Erasmus Law Rev 2(2):197–219

Giddens A (1990) The consequences of modernity. Stanford University Press, Stanford

Giddens A (1998) Risk society, the context of British politics. Polity, Cambridge

Giddens A, Beck U, Lash S (1994) Living in post-traditional society. Polity, Cambridge

Luhmann N (1993) Risk: a sociological theory. Walter de Gruyter, New York

Lupton D (1999) Risk. Psychology Press, Hove

Polman P (2014) Business, society and the future of capitalism. McKinsey Q, May issue

Praet M, Du Puy T (2016) Cultural diversity and political unity. In: Gretschmann K, Schepers S (eds) Revolutionising EU innovation policy. PalgraveMacmillan, New York

Schepers S (2011) The role of Board and CEO in managing societal and political intangibles. J Corp Gov 11(5):551–559

Schepers S (2016) Collaborative governance: a promising method for innovation. In: Gretschmann K, Schepers S (eds) Revolutionising EU innovation policy. Palgrave MacMillan

Turner F (2014) Tal der Egomanen. Die Zeit 52(12):8–9

Where Digitalization Meets Sustainability: Opportunities and Challenges

Sezen Aksin-Sivrikaya and C.B. Bhattacharya

1 Introduction

The globalization of digitalization has given rise to a vast amount of new services in both public and private sectors. Digitalization helps people find a common platform to voice their problems, concerns, and connect with the rest of the world. This not only transforms the way business is conducted but also enables citizens worldwide access better services in many areas such as healthcare and education.

As environmental concerns are rising on the horizon, digitalization makes the idea of a 'shared economy' possible. Digitalization enables owners and renters come together through online platforms and companies to share cars, accommodations, bikes, household appliances, and more. Sharing might be a solution to overconsumption and has potential environmental benefits through efficient use of resources. Rather than permanently owning an asset, many people may prefer buying a service whenever needed. Through this shared economy, we observe a switch from product orientation to service orientation in many industries.

This switch challenges the very existence of traditional business models. Today, firms operate within well connected networks of various actors where firm and industry boundaries start to disappear. Even though the situation poses a challenge for businesses, there are still exciting opportunities in which businesses can harness new technology and rethink existing business models to create value and be more sustainable.

S. Aksin-Sivrikaya (✉)
Humboldt University, Berlin, Germany
e-mail: sezen.aksin-sivrikaya@esmt.org

C.B. Bhattacharya
ESMT, Berlin, Germany
e-mail: cb@esmt.org

© Springer International Publishing AG 2017
T. Osburg, C. Lohrmann (eds.), *Sustainability in a Digital World*, CSR,
Sustainability, Ethics & Governance, DOI 10.1007/978-3-319-54603-2_3

The aim of this chapter is to build on service dominant (S-D) logic and interorganizational governance models to propose a conceptual framework for a digitalized ecosystem of multiple stakeholders where value is co-created. We would like to further identify potential emerging sustainable governance models within the proposed ecosystem.

The rest of the chapter is organized as follows. Section 2 briefly summarizes how technology has enabled us reap the benefits of digitalization and has been revolutionizing the way business is conducted through the idea of a shared economy. Section 3 presents how service dominant logic is a fit for current digitalized network structure. A network value co-creation model is proposed in Sect. 4, which is followed by emerging sustainable governance models in such business ecosystems in Sect. 5. Finally, Sect. 6 presents some challenges ahead and Sect. 7 concludes the chapter with our final thoughts.

2 How We Got Here

The move from the analog world to a digital one in the twenty-first century has brought increased processing power and communication speed, which facilitates information and data sharing. In this era of digitalization, stakeholders are greatly empowered through vast amount of information at their disposal. Amplified information availability not only assists stakeholders in learning about available product features and service offerings in the market, but also sharing goods and services. It was once common among friends and family to share things, but with digitalization, a community practice has become a profitable business model that initiates lower consumption, efficient use of resources, increased flexibility, and hence, a more sustainable society.

Previously non-digital products, such as bikes, watches, household appliances, become digitalized, which gave rise to a phenomenon called the 'Internet of Things' (IoT) (Atzori et al. 2010). Digitalization of every-day objects and their interconnection through IoT impacts the nature of traditional products and service offerings. GPS technology enables users locate goods and services within their vicinity in real time whenever needed. Payment systems further facilitate the use of these transactions by building trust in the system through intelligent e-commerce and invoicing systems (Black and Lynch 2004). In this way, hyper-efficient market places are created (Nov et al. 2010). The Internet facilitates aggregating supply and demand, where unmet demand is served and underutilized supply potential is unleashed. Social networks facilitate the matching process that brings supply and demand sides together (Constantinides and Fountain 2008).

Durable assets that sit idle most of the time can be utilized through sharing. Sharing enables people earn a rent by ownership and on the other hand, removes the ties of ownership and let people be independent by non-ownership (Chui et al. 2012). Luxury items that have been previously inaccessible to many consumers are now affordable. The rise of the shared economy is also supported by investors, as it is possible for companies to reap the benefits of new customer segments and markets. A total of $2 billion was invested in 200 sharing start-ups as of 2010 (Kriston et al. 2010; Chui et al. 2012).

Increased interconnectedness and information flow introduce a shift of power from centralized big companies to multi-stakeholder networks. As the sharing economy offers new value propositions by new market entrants and dramatically alters the environment in which firms function, firms need to respond with adequate business models that take these new challenges and threats into account to stay in business.

3   The Shared Economy and the Emergence of a Service-Dominant (S-D) Logic

This new logic of a shared economy identifies service provision as fundamental to economic exchange rather than the manufactured output which beckons us to introduce the concept of S-D logic. Vargo and Lusch (2004) define services as "the application of specialized competences (knowledge and skills) through deeds, processes, and performances for the benefit of another entity or the entity itself." Therefore, services cannot be treated as a residual or an add-on to the product offering anymore.

S-D logic introduces an ecosystem that is the operant resource and lever of competitive advantage. In this framework, the customer is not solely the receiving party anymore but also a partner as co-creator of value. While the incentive to participate in this network can be both monetary and non-monetary, collaboration is essential among participants as value creation relies on value exchanges between participants (Lusch and Nambisan 2015).

Within S-D logic, since value is co-created, value actualization is only realized if the customer accepts the value proposition made by the focal organization. If this is a durable product, customer interacts with the product itself as well as the service created in that process. Of course, producer value chain is different from a customer value chain. Manufacturers are generally quite objective in controlling technical qualities of a product but customers, on the other hand, use the product in their own individual way and experience its value differently (Gummesson 2008).

We argue that, in the era of digitalization, not only the customer but also other stakeholders such as business partners, suppliers, competitors, governments and NGOs become the co-creators of value. Each of these stakeholders may also have their own interpretations of value. This co-creation activity among multiple stakeholders and subjective value realizations transform the way value chains function in a shared economy.

S-D logic in the context of digitalization captures new market logistics spurred by IoT. S-D logic necessitates us to think in terms of network centricity rather than firm centricity, which is also the main feature of IoT. Of course, a network view complicates things as there are multiple interactions among many parties and companies usually fail to capture this level of complexity. In order to respond to this level of complexity with adequate business models, firms need to understand the network structure that digitalization imposes and how this structure changes value chains and business environment in return. In light of this, next section proposes a network model of value co-creation in the era of digitalization.

4 A Network Model of Value Co-creation in a Digital World

IoT is not as simple as a technology platform but rather a business ecosystem. We need to identify keystones of IoT business ecosystems, yet it is too early to tell which evolving ecosystems will be important or which players will become central. These players could be anyone, such as a device supplier, a supplier of software infrastructure, a supplier of hosted solutions or smart services, an IoT operator, a user community and many more. Therefore, instead of focusing on the players, it makes more sense to focus on the generation and capture of value in the ecosystems (Carbone 2009).

This transition requires a transformation from single firm oriented business models to ecosystem business models, focusing not only on a single firm's method of creating and capturing value but also value created and captured by other parts of the ecosystem (Westerlund et al. 2014). Being connected to other actors through technical and business ties in the ecosystem increases the level of complexity of the environment in which firms operate.

Existing business models are good at exploring single organizations but are not adequate when analyzing the interdependent nature of growth and success of companies that are evolving in the same ecosystem (Weiller and Neely 2013). In these ecosystems, information content is also very high, adding to the complexity. In such an environment, performance is highly dependent on collaborative competences, dynamic capability of customer orientation, and knowledge interfaces that facilitate innovative outcomes (Lusch and Nambisan 2015).

As the digital era moves away from goods-dominant logic to a service-dominant logic, there are implications for innovation processes, too. In S-D logic all innovations are service innovation; there is no longer the divide between product and service innovation. Due to its collaborative nature, there is a switch from the features and attributes of innovation output to the value that is co-created. Digitalization unleashes previously unused resources, and resource integration in the ecosystem becomes the way to innovate. Hence, innovation is not developed within firm boundaries anymore but by a network of actors (Lusch and Nambisan 2015).

Network effects are present when the value of goods and services increase with more consumers using them. As more and more goods and services become digital

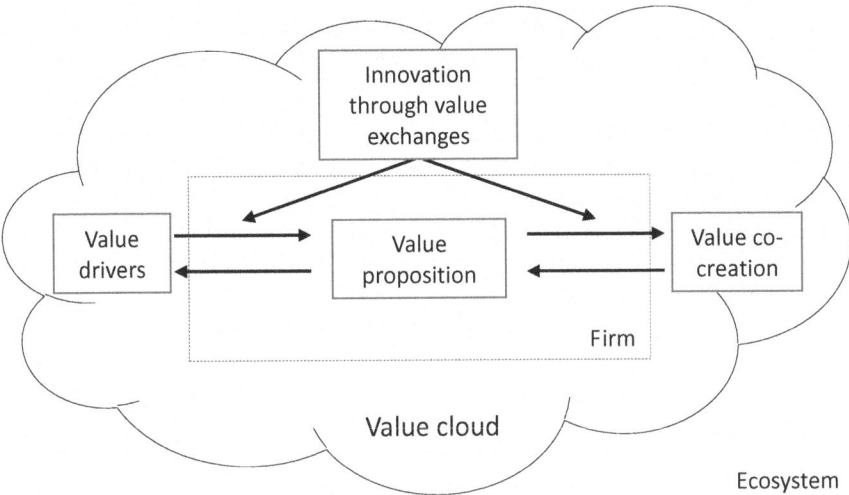

Fig. 1 A network model of value co-creation

and connected, network effects become the driver of value creation and key differentiator among competitors (Bharadwaj et al. 2013).

Figure 1 portrays our view of a network model of co-creation that digitalization brings upon us. It demonstrates the switch from a firm-centric view to a network-centric view where firm and industry boundaries almost disappear. The ecosystem we propose is composed of various actors, activities, or processes that are linked to generate value.

Value drivers can both be individual or shared motivations of different partici-pants and initiate an ecosystem to fulfill a need to innovate and create value. One can list key value drivers as sustainability, cybersecurity, and improved customer experience among others. There are ongoing interactions in the ecosystem, and through these, value drivers and value propositions are formed and reformed again continuously.

Firms and other actors in the system innovate through value exchanges, which are the exchange of value by different means, resources, knowledge, and informa-tion. Then, as a result of this process, value is co-created by multiple parties everywhere and anywhere in the ecosystem. Here, one cannot really think of a traditional value chain anymore but a 'value cloud'.

We argue that this model will ultimately become the norm not only for service firms but also for manufacturing firms as new market conditions move us towards a more service-oriented, sharing-based economy. In this environment, firms need to analyze the market, identify existing and entering actors and get ahead of trends to gain on competitive advantage.

It is also interesting to understand the sustainability implications of such eco-systems. This is what we look at next.

5   Strategy, Sustainability and Interorganizational Governance Models in the Era of Digitalization

Traditionally, business strategy has dictated IT strategy in firms. With digitalization though, business infrastructure facilitates increased interconnections between products, processes, and services. Digital strategy now transcends traditional functional areas such as marketing, logistics, procurement, finance, and HR (Bharadwaj et al. 2013). More importantly, though, digital strategy goes well beyond the limits of business to include environmental and social issues.

Digital strategy is "more than just bits and bytes, the digital infrastructure consists of institutions, practices, and protocols that together organize and deliver the increasing power of digital technology to business and society" (Hagel et al. 2011, p. 2). Therefore, it is not possible to decouple digital trends from sustainability trends anymore. We believe doing business through the sustainability lens empowered by digitalization will enable companies create value for the business, society, and planet.

In the digital era, digital strategy and sustainability strategy will become integral parts of corporate strategy. Corporate strategy in this era will rely on rich information exchanges among multiple parties and extend supply chains to dynamic ecosystems that go beyond traditional firm and industry boundaries as we have argued so far.

As a consequence of this evolution in corporate strategy, new governance models are bound to develop. In order to make predictions about the nature of these models we will transition from the existing literature on interorganizational governance models in networks to network governance models which have emerged because of the need to understand dependence relationships among multiple stakeholders in a network. In the latter type of networks, there is usually a high level of complexity and uncertainty where the degree of connectedness and ease of information flow among participants vary (Powell et al. 1996; Jones et al. 1997).

These models that are based on extensive collaboration and joint decision making are well established in the literature (Jones et al. 1997; Dyer and Singh 1998). There are also extensions to sustainable supply chain management that underline the benefits of combining social and environmental issues along the supply chain. In such models, due to social pressure, collaboration becomes essential as participants are forced to seek multilateral benefits at the network level rather than unilateral benefits at the firm level (Carter 2000; Gereffi et al. 2005; Drake and Schlachter 2008; Vurro et al. 2009).

In these studies, success factors for organizations are identified as the ability to apply integrated approaches based on long term cooperation, knowledge exchanges, and joint upstream and downstream competence building (Maignan et al. 2002; Strand 2009). Even though digitalization has been changing the dynamics of the value chain as we have argued in the previous section, we believe that long term cooperation, knowledge exchanges, and joint competence building are also key success factors in achieving sustainable digital ecosystems. Of course, firms differ in their approaches to collaboration and sustainability (Roberts 2003; Jiang 2009), which, in turn, affect the nature of these newly emerging governance models.

Our main goal here is to take network governance models a step further and adapt them to digital business ecosystems. Specifically, we would like to explore the interplay between these models and conditions under which sustainability is successfully embedded into the activities and processes that take place within the ecosystem.

Network Structure and Governance Models for Sustainability

Corporate responsiveness to sustainability pressures in stakeholder networks is dependent on two key features. The first one is *network density*, which is the degree of completeness of the ties between the participants in a network. When participants in a network are better connected, information flow is more efficient, which forces organizations to be more responsive (Meyer and Rowan 1977; Oliver 1991). The second feature in our framework is what we will call *organizational influence*. Also called *centrality* in the governance literature, it reflects the extent of a firm's relative power or status in a network (Brass 1984). It also refers to the firm's ability to control flow of information and act as a gatekeeper (Bonacich 1972; Freeman 1978).

We argue that these two features will also be the determinants of the degree of sustainability embeddedness in the digital business ecosystem. The ecosystem has no geographical or industrial boundaries and the focal organization can be any operational stakeholder in the ecosystem. Figure 2 illustrates each combination of these two features and our expected outcomes, which are further explained next.

Dictatorial sustainability—Dictatorial sustainability is observed in more traditional, less digitized environments where there is one powerful organization that exerts influence across a low density ecosystem. This organization could be a firm,

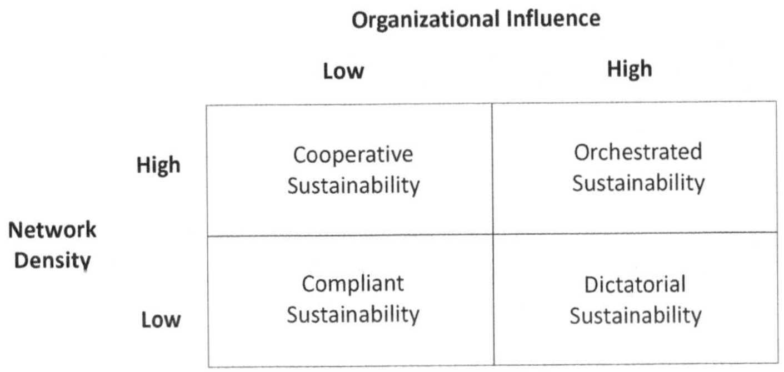

Organizational Influence

		Low	High
Network Density	**High**	Cooperative Sustainability	Orchestrated Sustainability
	Low	Compliant Sustainability	Dictatorial Sustainability

Fig. 2 Different combinations of the key features of corporate responsiveness to sustainability pressures in stakeholder networks

an influential community, the government or even an NGO. The organization can either resist pressures from others to conform to sustainability expectations or impose self-centered practices that reflect its own interpretation of sustainability.

Compliant sustainability—Similar to *dictatorial sustainability*, we observe *compliant sustainability* in less digitized, potentially traditional environments where information flow is limited and there are potentially too many players widely dispersed in the business ecosystem. *Compliant sustainability* occurs when organizations lack influence and network density is low. In such an environment, there is not an incentive to integrate sustainability as no party is powerful enough to exert influence across the network and push participants for sustainability. In this quadrant, we observe ad hoc implementation of social and environmental initiatives to temporarily meet the demands of threatening stakeholders, especially regulators.

Cooperative sustainability—We observe *cooperative sustainability* when the ecosystem is decentralized and organizations lack influence in a dense network. Network density facilitates information flow and firms feel stakeholder pressures. This forces organizations to conform, compromise, and bargain with other stakeholders in order to remain in the ecosystem. In these types of ecosystems, multistakeholder collaborations emerge to develop joint frameworks to achieve sustainability.

Orchestrated sustainability—*Orchestrated sustainability* arises in business ecosystems with high density and a powerful organization at the center. Focal organization responds to the concerns of the ecosystem and adopts sustainable business practices. Due to its influence it becomes the leader and oversees collaboration in the ecosystem. It guides other stakeholders in adopting sustainable practices and processes by facilitating establishment of certification programs, knowledge sharing platforms, and sustainable management schemes.

In the long-run, we expect that the network density will reach to a point where the bottom-left and bottom-right corners in Fig. 2 will not be attainable even for traditional business environments anymore. With increased integration of digitalization into our lives and businesses, we anticipate the emergence of *orchestrated sustainability* or *cooperative sustainability*. It will not be possible for firms to impose their own values or get away with being only compliant in a digital business ecosystem. Emergence of these two governance models will facilitate creating a shared culture among partners to benefit from relational rents, stimulate innovation processes, and improve adaptability to ever changing business environment in the era of digitalization.

6 Challenges Ahead

Due to its scope, this transformation will bring many challenges in different venues. We decompose the potential challenges into three categories in this section: challenges of designing business models in a digital environment, disruptions to existing businesses, and environmental and societal issues caused by digitalization.

Designing Business Models in a Digital World

First of all, there are technical and operational difficulties that come with digitalization. Business model design will not be easy due to the immaturity of products and services and the level of complexity that the network structure introduces.

To start with, there are too many different types of connected objects with only modestly standardized interfaces. There are countless ways of connecting an object, a business, and a consumer together, which creates endless possible business models (Leminen et al. 2012). There are presently 10 billion devices connected and this number is estimated to be 50 billion by 2020. As a matter of fact, more than 99% of physical objects that may become part of the network are not yet connected (Evans 2011).

IoT technologies are not yet standard products and services; they are quite immature and complex. Ecosystems are therefore currently unstructured; it is too early to tell who the actual participants are and which roles they will have. It is hard to define the underlying structures, governance, and specific value creating logics (Westerlund et al. 2014).

Customer demands such as flexibility, high quality at a small cost, and superior experience add to the complexity, which requires firms to collaborate across networks. As the business systems get complicated, customer interfaces have to be kept simple and intuitive. Furthermore, while serving different geographical regions, companies need to adapt and learn to deal with different cultures and languages (Prahalad and Krishnan 2008). Some players such as Uber and Airbnb are criticized in certain regions since they do not really fit standard customer and regulator expectations. In order to meet these expectations, companies need to work closely with regulators among other local actors in respective regions.

Networks also necessitate acquiring resources globally. Firms need to use specialized suppliers; vertical integration is not possible anymore. Speed is even more important now as companies try to gain on scalability as well as serving their customers using resources coming from all over the world under competitive pressures (Prahalad and Krishnan 2008). The ability to orchestrate the supply chain is a source of competitive advantage. This requires working in a collaborative fashion from conceptual design to recycling of products by dynamic realignment of partners and suppliers along the supply chain (Bharadwaj et al. 2013).

Digital Disruption

Within business ecosystems, cooperation will be initially difficult to achieve as some traditional players will cease to exist as a result of the technological shift that digitalization imposes. Digital attackers might disrupt existing business models beyond country borders. For instance, Snapchat made mainstream media look obsolete, operating on a platform-as-a-service infrastructure. Similarly, Simple challenged big-cap banks, without even having a single branch (Dawson et al. 2016).

Brand new value propositions cause huge shifts in markets by introducing goods and services that customers were not aware that they needed in the first place. For instance, Amazon and many others transformed storage into a service and hence traditional business models of hard-drive makers became less relevant. This and many other examples in this realm change how value chains function by reducing fixed and variable costs and turning products into services (Dawson et al. 2016).

Companies such as Google, Facebook or Amazon take advantage of improvements in computer hardware, software, and connectivity when developing and launching products (Bharadwaj et al. 2013). They also enjoy network benefits of serving millions of customers. Their operational leverage enables them to upsell or cross-sell products with no or limited human interaction, which bring in substantial financial advantages. Such platforms also create barriers to entry by forcing the rest to integrate into an ecosystem built by the platform (Dawson et al. 2016). These companies also make use of multisided business models in which they offer free products or services in one layer to capture value at a different layer. Value creation through coordinated business models in networks is also possible. Content providers can coordinate and time their offerings to co-create value. Some companies such as Apple appropriate value through control of digital industry architecture. Under these conditions traditional companies find their capabilities misaligned and themselves at a competitive disadvantage (Bharadwaj et al. 2013).

Environmental and Societal Implications

Digitalization has also significant implications for sustainability. As a result of digitalization, we are bound to use electronic devices. Electronic waste or e-waste can be described as the discarded electronic devices. If not done right, processing e-waste can lead to adverse health effects and environmental pollution. Electronic scrap components contain harmful materials such as lead, cadmium, and brominated flame retardants. E-waste is usually exported to developing countries and processed under less than ideal conditions. However, great care has to be taken to prevent unsafe exposure in recycling operations and leakages from landfills and incinerator ashes (Sthiannopkao and Wong 2013).

Exchange of goods and services in a digital environment also poses a threat on the security and privacy of the involved parties in transactions. For instance, in the insurance sector, the fact that people's eating habits and exercise patterns can be monitored through wearables may let companies earn premiums for poor eating or exercise habits. Currently, there are different legal data protection requirements across countries and there is need for a general legal framework for access to personal information. This could be done by an international legislator guided by the private sector (Weber 2010).

On the other hand, digitalization has had dire consequences in the realm of human rights, in relation to the minerals that are used in electronic devices. Revenues from conflict minerals that are extracted from the mines at the Democratic Republic of the Congo fuel civil war in the region. Extraction has cost millions of kids their future and civil war resulted in deaths of millions. There have been international efforts to stop trading activities from conflict smelters. For instance, in the US, Dodd-Frank Wall Street Reform and Consumer Protection Act of 2010, required firms to audit their supply chains and report use of conflict minerals. We need to extend these efforts and enforce auditing requirements worldwide, since this issue will keep escalating with digitalization and the uptake of IoT.

7 Final Thoughts

The new era will be marked by collaborative behavior, social networks, and professional and technical workforce. It will be the end of industrious thinking of immense commercial activity and mass labor forces. Technological advancements such as 3D printing will let individuals become manufacturers themselves, which has the potential to make highly capitalized, centralized factories obsolete (Rifkin 2011).

We will observe more and more integrated business ecosystems, while firm-centric views will cease to exist. Firm boundaries will fade and value creation activities will increasingly take place within a network of various actors who will co-create value together.

The new era will not come without its challenges, though, in particular those related to the design of business models in a digital world, disruption to existing businesses, environmental and societal concerns. Once we learn how to deal with these issues, emerging sustainable governance models will reduce frictions and costs associated with information collection and processing, management, energy use, manufacturing, and logistics. This, in turn, will dramatically and irreversibly change the way of doing business not only for the benefit of the business itself but also for the environment and society at large.

References

Atzori L, Iera A, Morabito G (2010) The internet of things: a survey. Comput Netw 54(15): 2787–2805

Bharadwaj A, El Sawy OA, Pavlou PA, Venkatraman NV (2013) Digital business strategy: toward a next generation of insights. Mis Q 37(2):471–482

Black SE, Lynch LM (2004) What's driving the new economy?: the benefits of workplace innovation. Econ J 114(493):F97–116

Bonacich P (1972) Factoring and weighting approaches to status scores and clique identification. J Math Sociol 2(1):113–120

Brass DJ (1984) Being in the right place: a structural analysis of individual influence in an organization. Admin Sci Q 1:518–539

Carbone P (2009) The emerging promise of business ecosystems. Open Source Business Resource, Feb 2009. http://timreview.ca/article/227

Carter CR (2000) Ethical issues in international buyer–supplier relationships: a dyadic exami-
    nation. J Oper Manage 18(2):191–208
Chui M, Manyika J, Bughin J, Dobbs R, Roxburgh C, Sarrazin H, Sands G, Westergren M (2012)
    The social economy: unlocking value and productivity through social technologies.
    McKinsey Global Institute
Constantinides E, Fountain SJ (2008) Web 2.0: conceptual foundations and marketing issues.
    J Direct Data Digit Mark Pract 9(3):231–244
Dawson A, Hirt M, Scanlan J (2016) The economic essentials of digital strategy. McKinsey Q
Drake MJ, Schlachter JT (2008) A virtue-ethics analysis of supply chain collaboration. J Bus
    Ethics 82(4):851–864
Dyer JH, Singh H (1998) The relational view: cooperative strategy and sources of inter-
    organizational competitive advantage. Acad Manage Rev 23(4):660–679
Evans D (2011) The internet of things—how the next evolution of the internet is changing every-
    thing. Cisco Internet Business Solutions Group (IBSG) White Paper.
Freeman LC (1978) Centrality in social networks conceptual clarification. Soc Netw 1(3):215–239
Gereffi G, Humphrey J, Sturgeon T (2005) The governance of global value chains. Rev Int Polit
    Econ 12(1):78–104
Gummesson E (2008) Extending the service-dominant logic: from customer centricity to
    balanced centricity. J Acad Mark Sci 36(1):15–17
Hagel III J, Brown JS, Kulasooriya D (2011) The 2011 shift index: measuring the impact of long-
    term change. Deloitt Center for the Edge Report, Deloitt Development LLC 2011
Jiang B (2009) Implementing supplier codes of conduct in global supply chains: process expla-
    nations from theoretic and empirical perspectives. J Bus Ethics 85(1):77–92
Jones C, Hesterly WS, Borgatti SP (1997) A general theory of network governance:
    exchange conditions and social mechanisms. Acad Manage Rev 22(4):911–945
Kriston A, Szabó T, Inzelt G (2010) The marriage of car sharing and hydrogen economy:
    a possible solution to the main problems of urban living. Int J Hydrog Energy 35(23):
    12697–12708
Leminen S, Westerlund M, Rajahonka M, Siuruainen R (2012) Towards iot ecosystems and
    business models. In: Andreev S, Balandin S, Koucheryavy Y (eds) Internet of things,
    smart spaces, and next generation networking. Springer, Berlin, pp 15–26
Lusch RF, Nambisan S (2015) Service innovation: a service-dominant logic perspective. Mis Q
    39(1):155–175
Maignan I, Hillebrand B, McAlister D (2002) Managing socially-responsible buying: how to
    integrate non-economic criteria into the purchasing process. Eur Manage J 20(6):641–648
Meyer JW, Rowan B (1977) Institutionalized organizations: formal structure as myth and cere-
    mony. Am J Sociol 1:340–363
Nov O, Naaman M, Ye C (2010) Analysis of participation in an online photo-sharing community:
    a multidimensional perspective. J Am Soc Inf Sci Technol 61(3):555–566
Oliver C (1991) Strategic responses to institutional processes. Acad Manage Rev 16(1):145–179
Powell WW, Koput KW, Smith-Doerr L (1996) Interorganizational collaboration and the locus of
    innovation: networks of learning in biotechnology. Admin Sci Q 1:116–145
Prahalad CK, Krishnan MS (2008) The new age of innovation. McGraw Hill, New York
Rifkin J (2011) The third industrial revolution: how lateral power is transforming energy, the eco-
    nomy, and the world. Macmillan, London
Roberts S (2003) Supply chain specific? Understanding the patchy success of ethical sourcing
    initiatives. J Bus Ethics 44(2-3):159–170
Sthiannopkao S, Wong MH (2013) Handling e-waste in developed and developing countries:
    initiatives, practices, and consequences. Sci Total Environ 463:1147–1153
Strand R (2009) Corporate responsibility in Scandinavian supply chains. J Bus Ethics 85(1):
    179–185
Vargo SL, Lusch RF (2004) Evolving to a new dominant logic for marketing. J Mark 68(1):1–7

Vurro C, Russo A, Perrini F (2009) Shaping sustainable value chains: network determinants of supply chain governance models. J Bus Ethics 90(4):607–621

Weber RH (2010) Internet of things–new security and privacy challenges. Comput Law Secur Rev 26(1):23–30

Weiller C, Neely A (2013) Business model design in an ecosystem context. University of Cambridge, Cambridge Service Alliance, Cambridge

Westerlund M, Leminen S, Rajahonka M (2014) Designing business models for the internet of things. Technol Innov Manage Rev 4(7):5

Leadership in a Digital World: New Ways of Leadership for Sustainable Development

Christiane Lohrmann

1 Introduction

We are in the middle of a digital revolution that will fundamentally alter the way we live, work and relate to one another. In its scale and complexity, this transformation will be unlike anything we have experienced before. We know already that the ways in which we interact are changing, becoming ever more integrated and comprehensive, and involving all stakeholders, from the public and private sectors to academia and civil society. This implies an immense shift in the way we understand leadership of ourselves, our teams and entire organisations.

2 The Future of Work in a Digital World

We are already witnessing an incredibly rapid change in the way we work (Hay Group 2015a, b). According to new OECD research, in Germany alone, it is expected that 12% of all jobs will disappear due to automation (OECD 2016). Digitalisation is indeed changing the world of work and the ways in which we work together. Exponentially increasing computing power, Big Data, the penetration of the Internet, artificial intelligence (AI), the Internet of Things and online platforms are among the developments which are radically changing prospects for the types of jobs that will be needed both today and in the future. According to the OECD study,

---

'The finished work gets admired, the work in progress gets underrated'. Friedrich Nietzsche (German philosopher, 1844–1900)

C. Lohrmann (✉)
FranklinCovey Leadership Institut, München, Germany
e-mail: post@christianelohrmann.de

© Springer International Publishing AG 2017
T. Osburg, C. Lohrmann (eds.), *Sustainability in a Digital World*, CSR, Sustainability, Ethics & Governance, DOI 10.1007/978-3-319-54603-2_4

'Automation and independent work in a digital economy', there are four main developments in the field of work in the digital economy:

- Digitalisation is reducing demand for routing and manual tasks while increasing demand for **low- and high-skilled tasks and problem-solving and inter-personal skills**.
- Digitalisation has opened the door to **new forms of work organisations**. Although the 'platform economy' may be efficient in matching workers to jobs and tasks, it also raises questions about wages, labour rights and access to social protection for the workers involved.
- Digitalisation raises questions about technology's potential to replace workers. Estimates based on the Survey of Adult Skills (PIAAC) show that, on average and across countries, 9% of jobs are at high risk of becoming automated, while for another 25% of jobs, **50% of the tasks involved will change significantly because of automation**.
- Digitalisation will provide **new opportunities** for many but will present challenges for others, including the risk of growing inequality with respect to access to jobs and their quality as well as career potential. We need more rather than fewer politics to allow workers to grasp new opportunities and respond to challenges.

It is obvious that digitalisation is leading to new opportunities for improving our lives in the future. It is also clear that we are in need of self-management as well as leadership skills to solve problems and meet challenges in rapidly changing and innovative work organisations. As OECD calculations show (see Fig. 1), mostly high-skilled and medium, non-routine jobs have increased in the past 12 years, and thus it is here where leadership skills are most highly needed (OECD 2016).

Fig. 1 Job polarisation in the European Union, Japan and the United States (OECD 2016)

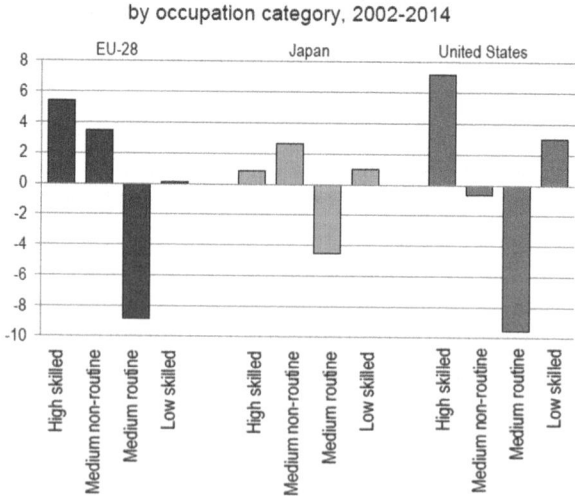

Percentage-point change in employment shares by occupation category, 2002-2014

3 Leadership and Culture in Change: Five Types of 'Good Managers'

What does this development mean for management and leadership in a digital world? Constant change and the need for innovation is required. For leaders and managers, this means that enhancing resilience as well as the ability of people and systems to cope with change should have the highest strategic priority. More than ever, it is important to motivate people, help them become effective and productive employees in digital work environment, and unleash their talents. Therefore, curiosity, a readiness to take risks, and the capacity to deal with uncertainty are becoming increasingly important. The greater the networking is within a company, the better that company can deal with changing circumstances. After all, network building in and between companies is the best response to the challenges of the modern working environment. This implies that managers must evolve into effective leaders who can coach and guide their teams and, eventually, the whole organisation (Goldsmith, in Simpson 2014, p. 3). So, the aim is less to exercise management via goal setting and controlling and more to design the best possible framework for collaboration, which will in turn unleash talent and network building. Here, relinquishing power is just as important as the transparency of information. Despite the high degree of autonomy enjoyed by the stakeholders' networks, common agreement on basic values and a consideration of contexts will enable a much needed alignment of thoughts, systems and actions. Common principles and values are the foundation upon which successful company performance and a strong company culture are built. Today, personal satisfaction, mutual recognition and reputation have more influence on motivation than financial incentives; likewise, self-determination has become more precious than status symbols. Moreover, the contemporary public discusses and evaluates not just the economic performance of companies but their role as actors in society as well. After profit margins and shareholder value, reputation and company culture have become key benchmarks of good management.

In this context, leadership obtains a new role. The 2014 Initiative Neue Qualität der Arbeit (INQA) study, '*Führungskultur im Wandel*', claims that more than two-thirds of all leaders interviewed were not satisfied with leadership culture as it exists today. According to INQA, one of Germany's leading think tanks regarding the future and quality of work, leaders are starting to realise the need for new leadership mindsets and skills (Sattelberger 2016). Apart from a focus on open processes, which was mentioned by 100% of the managers interviewed (INQA 2014), individual differences in the semantic mapping of 'good management' demonstrated five types of manager preferences in Germany.

Type 1 'Traditional Care and Reassurance' (13.50%)
According to the study, a type 1 manager has the ability to give people a feeling of security and personal reassurance. For this type, good management is authentic, competent and endowed with natural authority. Loyalty and satisfaction on the part

of company employees are the outcome of a personal role model function and assumption of responsibility on the part of the manager. The chief aim of this type of manager is to secure peoples' jobs in the company in the long term.

Type 2 'Profit-Enhancing Management' (29.25%)

This type of manager is able to organise people in such a way that they can extract maximum profit from an existing business model. Good management enhances the competitive edge of the company through strategy, goal setting and professional, KPI-driven controlling. The chief aim of type 2 managers is to secure attractive profits for shareholders.

Type 3 'Coaching for Cooperative Teamwork' (17.75%)

Type 3 managers support and supervise cooperation in decentralised teams, which can easily adapt to multiple tasks. Good management promotes in-house diversity, ensures maximum transparency, and enables discourse and joint reflection on contexts and interrelations. The chief aim of type 2 managers is to leverage synergies both inside the company and between the company and other stakeholders.

Type 4 'Stimulating Network Dynamics' (24.00%)

Type 4 managers provide leeway for personal initiative and encourage the unimpeded, non-hierarchical networking of the whole cast of players in the company. Good management unites people of diverse backgrounds and lifestyles into one appealing vision and trusts their ability to self-organise. The chief aim of type 4 managers is to create internal networks as complex as their external counterparts.

Type 5 'Acting in Solidarity with Stakeholders' (15.50%)

Finally, Type 5 managers primarily motivate through personal appreciation, self-determination and the meaningfulness of shared working experience. 'Good management' here is open to grass-roots democracy, while social solidarity and social responsibility are important and heavily emphasised themes in its day-to-day dealings. The chief aim of type 5 managers is to balance the interests of all relevant stakeholders.

4 Why Are Leadership and Coaching Sustainable Solutions for our Digital World?

As mentioned above, we are witnessing a paradigm shift whereby people are increasingly viewing reputation and company culture as the key benchmarks of good management as opposed to profit margins and shareholder value. Therefore, company culture is becoming the ultimate advantage of organisations. After all, company culture is created by the behaviour of its leaders. According to management expert Stephen R. Covey, leadership and management are two different things: 'Leadership is not management. Leadership has to come first. Management is a bottom-line focus: How can I best accomplish certain things? Leadership deals with the top line: What are the things I want to accomplish?' (Covey 2004, p. 101). And, in the words of Peter Drucker and colleagues, 'Management is doing things right; leadership is doing the right things. Management is efficiency in climbing the ladder of success; leadership determines whether the ladder is leaning against the right wall' (Drucker 2006, p. 3).

Leaders who transform their lives, their teams, and their organisations model the highest levels of personal and interpersonal effectiveness and are able to achieve long-lasting and sustainable results for their organisation (Franklin Covey Leadership Institut 2017).

Organisations therefore need to develop leaders at three levels: personal, team and organisation. At every level, two aspects should be discussed:

(a) Character—how can leaders be an example of personal effectiveness, build trust with all those who are involved, and increase their circle of influence?
(b) Competence—how can leaders motivate others to effectively set goals to achieve lasting and certain success?

According to Covey (2004, 2008), there are four principles for leadership (see Fig. 2):

1. The first is to inspire trust. You build relationships of trust through both your character and competence, and you also extend that trust to others. You show others that you believe in their capacity to live up to certain expectations, to deliver on promises, and to achieve clarity on key goals. You do not inspire trust by micromanaging and second guessing every step people make.
2. The second is to clarify purpose. Successful leaders involve their employees in the communication process to create the goals that need to be achieved. If people are involved in the process, they will psychologically own it, and a situation will be created whereby all relevant parties are on the same page about what is really important—mission, vision, values and goals.

Fig. 2 The four principles of leadership according to Covey (Franklin Covey Leadership Institut 2017)

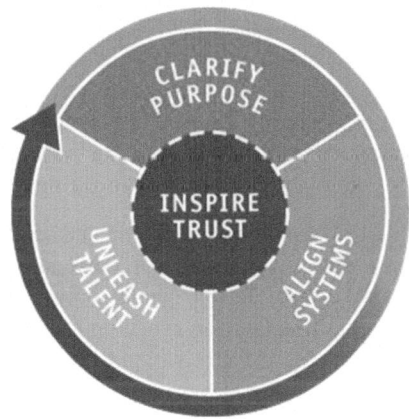

3. The third is to align systems. This means that you do not allow there to be conflict between what you say is important and what you measure. For instance, many times, organisations claim that people are important but, in fact, the structures and systems, including accounting, make them an expense or cost rather than an asset and the most significant resource.

4. The fourth is the fruit of the other three—unleashed talent. When you inspire trust and share a common purpose with aligned systems, you empower people. Their talent is thereby unleashed so that their capacity, intelligence, creativity and resourcefulness can be utilised. (Covey 2017)

In addition to leadership skills, coaching as a means to lead, guide, support, motivate and inspire people in a digital world is becoming ever more important in the contemporary world (Simpson 2014, p. 2). Coaching is about people. However, coaching is more than just consulting or advising; it is a specific set of competencies, skills and behaviours, and it requires a certain kind of good intent and character. Coaching is about building a relationship of trust, tapping a person's potential, creating commitment and executing goals.

According to Michael Simpson (2014, 3f), there are four principles of coaching:

(a) Trust

Trust is hard to earn but easy to lose. It can take weeks or months of careful nurturing to cultivate trust—whereas one broken promise, one display of indifference, one manipulation with bad intent, or one breach of confidence can ruin everything. This is why all effective coaching starts with an understanding of the great obligation to be trustworthy (Simpson 2014, p. 15).

(b) Potential

Coaches can help people recognise their potential instead of their limitations. They can help fuel, support and spark imagination and talent (Simpson 2014, p. 24).

(c) Commitment

Creating lasting commitment is a key principle of effective coaching, and the principal skill for creating commitment is to ask powerful questions (Simpson 2014, p. 30).

(d) Execution

Coaching is working to discover the precise nature of an individual's desired destination. The task is to help people execute their commitments and be held accountable. Moreover, the best coaches can actually help individuals get into a 'flow' state that can be inspiring for them (Csikszentamihalyi 2009).

Efforts to enact organisational change are never easy. However, if done right, the rewards for sustainable solutions and long-lasting success are immense. After all, the rewards of change can be more important than money or profit margins. In this context, leadership and coaching become important tools for fostering engagement, involvement and motivation from employees. They make sure that each person in the organisation knows his or her part in the process and play a key role in the transformational effort of our digital world.

5   Conclusion

Our world is changing rapidly due to digitalisation. As a consequence, new forms of collaboration and organisations are evolving. While many jobs may disappear, it is clear that leadership skills will be needed more than ever; so that in this respect, polarisation is to be expected. There is the risk of growing inequality in terms of access to jobs and their quality, as well as career potential. We need more rather than less policies and principles to allow workers to grasp new opportunities and respond to new challenges.

However, we know that innovation comes not only from technology but to a great extent from the ways in which we work together and arrange our organisations. This is why leadership and coaching skills will become more important than ever for the sustainable success of organisations. As already pointed out in recent INQA research, many company leaders today are aware that leadership skills need to change rapidly and substantially to keep up with technology. Among the challenges and newly required skills arising from this situation are the stimulation of network dynamics, openness towards stakeholder issues, motivation through personal appreciation, self-determination, the meaningfulness of shared working experience, and both supporting and supervising cooperation in decentralised teams that can easily adapt to multiple tasks. This article emphasises the importance of company culture in today's digital world as a competitive advantage to attract talent, motivate people and be sustainably successful. It also points out the difference between leadership and management while presenting important leadership and coaching principles, such as creating trust, clarifying purpose, aligning systems and unleashing talent and positive commitment from people and towards achieving company goals in a sustainable world.

References

Covey SR (2004) The 7 habits of highly effective people. Simon Schuster, New York

Covey SR (2008) The speed of trust. Simon Schuster, New York

Covey SR (2017) The leader formula: the 4 things that make a good leader. http://www.stephencovey.com/blog/?p=6. Accessed 6 Jan 2017

Csikszentamihalyi M (2009) Flow: the psychology of optimal experience. Harper Collins, New York

Drucker PF (2006) Managing for results. Harper Business, New York

Franklin Covey Leadership Institut (2017) Leadership. http://www.franklincovey.com/leadership/. Accessed 6 Jan 2017

Hay Group (2015a) Digitization and your workforce: why technology isn't going to change your workforce the way you might think—and what you can do about it. http://www.haygroup.com/en/our-library/whitepapers/digitization-and-your-workforce/#.WHCbcVxtbqA

Hay Group (2015b) Führungskräfte für eine neue Welt: Was die Zukunft von Führungskräften verlangt. http://www.haygroup.com/downloads/de/Leadership_2030_Whitepaper_DE_web.pdf. Accessed 6 Jan 2017

Initiative Neue Qualität der Arbeit (INQA) (2014) Führungskultur im Wandel, Kulturstudie mit 400 Tiefeninterview. https://www.inqa.de/SharedDocs/PDFs/DE/Publikationen/fuehrungskultur-im-wandel-monitor.pdf?__blob=publicationFile. Accessed 6 Jan 2017

Organisation of Economic Cooperation and Development (OECD) (2016) Automation and independent work in a digital economy, policy brief on the future of work. http://www.oecd.org/employment/Automation-and-independent-work-in-a-digital-economy-2016.pdf. Accessed 6 Jan 2017

Sattelberger T (2016) Digitalisierung & ihre Auswirkungen—auf die Arbeitswelt und auf jeden Einzelnen, Speech with Slides, Munich GLS Bank, 30 Nov 2016

Simpson M (2014) Unlocking potential, 7 skills that transform individuals, teams and organisations. Grand Harbor Press, Grand Haven, MI

Sovereign Decisions as a Means for Strengthening Our Resilience in a Digitalized World

Denise Feldner

1 Introduction

"Today, no country is cyber ready" claimed the Potomac Institute for Policy Studies in the opening statement of its Cyber Readiness Index 2.0 for international readers (Hathaway et al. 2015). Nevertheless, on July 25, 2015, the long-awaited and controversially discussed German IT Security Act (Gaycken 2015) came into force to improve the security of information technology systems in Germany (Gabel and Schuba 2015). The federal government wants Germany's IT systems and digital infrastructure to be among the most secure in the world. With the approval of this draft legislation in December 2014, the Federal government started the implementation of the "Digital Agenda 2014–2017". With the agenda, the country is leaning towards utilizing opportunities that digitalization presents to strengthen Germany's role in a global market.

In 2016, the German government launched the White Paper 2016 on German security policy (Bundesregierung 2016). It is a contribution by the German government to security debate in the country. It shows international partners how Germany defines its role in the world in terms of security policy. It replaced the White Paper 2006, the last of its kind. For the first time in German history of security policy the Federal Armed Forces will also focus on offensive measures to protect critical infrastructures, citizens' privacy, government institutions, and businesses from cyberattacks.

Enough to Get Done What Ought to Be Done to Be Cyber Ready?
It is a given that global economic growth is increasingly dependent upon the rapid adoption of new technologies. Digitalization, one of today's greatest challenges, equally relates to aspects of (I) security, (II) economy, and (III) society. While

D. Feldner (✉)
German U15 e.V., Berlin, Germany
e-mail: denise.feldner@german-u15.de

digitalization opens up important economic perspectives, it also confronts political and company leaders with new challenges caused by the global implementation of a disruptive innovation called the internet on a global scale. Given the increasing impact of cyberspace on everyday life, the German government has a particular interest in maintaining a peaceful, free, open and secure internet.

As a major stakeholder in the European Union, Germany focuses primarily on the European Single Market. However, this market with over 500 million consumers requires not only high common standards, but also a well-coordinated and cooperated cyber foreign and security policy based on contemporary norms. The internet, which largely defies traditional national borders, has produced new approaches of intergovernmental cooperation, multi-stakeholder regimes, multilateral bodies, new forms of cooperation as well as new forms of communication among citizens.

In addition, security policy is still changing in fundamental ways. Virtual attacks, information and cyber warfare threatening critical infrastructures, government institutions, and long-term partnerships such as the transatlantic partnership represent some of the strongest present-day challenges to security policy. A secure internet is essential to the protection of individual liberties, the right to informational self-determination of citizens and of democracy as a whole. In 2016 global anarchy in cyberspace, with all of its inherent perils, still persists.

Four central challenges for democratic governance emerge from this:

- The blurring of distinctions between internal and external policies
- Protectionism
- Privatization of governance
- New forms of cooperation and participation

In order to reach a broad audience, I will give an overview on the topic. This article does not include empirical data or analysis. Instead, I will refer to bibliographic resources and web references.

2 Security

Surveillance scandals and threats to democracy are on the rise, with the internet being the backbone of most spy programs and technologies. Moreover, breaches of trust that have already occurred will pose long-term challenges for international cooperation. For example, it was revealed that the U.S. had been hacking Israeli drones for years (Currier and Moltke 2016). But besides that, massive attacks against Georgia during its war with the Russian Federation in 2008, as well as cyber incidents with the Stuxnet attack on Iran's nuclear plant in 2010, have taken place. In 2016, the U.S. government officially accused Russia of orchestrating a hacking campaign to interfere with U.S. elections.

After several years of surveillance the U.S. government also announced in the same year it will launch a cyber warfare campaign on the Islamic State of Iraq and the Levant (Isil) to decrease its communication and marketing channels. In 2017, the U.S. launched a campaign called "left of launch" against North Korea to prevent missile strikes.

These events prove that evidence that states are focusing on the subject of cybersecurity through prevention, deterrence and attacks, driven by security or economic needs. Additionally, this led to deliberations on how to limit data traffic in such a way that it is out of reach of foreign countries' security officials.

At the same time cybercrime has become a business which exceeds a trillion dollars a year in online fraud, identity theft, and lost intellectual property. This affects millions of people around the world, as well as countless companies and the governments of nearly every nation. The loss is estimated to cost German companies an average of nearly 5 million euros per year (Bendiek 2014) to 22.3 billion per year (BITKOM 2015).

Safeguarding of The Internet Infrastructure

As a result, securing the internet infrastructure, data spaces and services has become of highest importance to both governments and company leaders at the same time. Germany finds itself in a distribution battle over the internet's possibilities to gain power over sovereign countries, their citizens' data, and over market shares in the internet as a global marketplace. This makes the topic a global issue for both security policies and economy. For instance, in its new 2016 National Security Strategy (Hammond 2016), the United Kingdom again characterized cyberattacks, including attacks by other states and by organized crime and terrorists as one of four "Tier One" threats to British national security (HM Government 2010). Russia published its cyber concept for the armed forces a second time in 2011 (Russian Federation 2011), stipulating that the role of the information war has grown substantially.

The current discussion is taking place in an environment of fundamentally differing starting points for an appropriate objective of cyberspace regulations. These differences over the right balance of interests are results of differing geo-strategic positions of the nations involved. On the one hand, this has to do with different views on privacy and personal rights of citizens, on the other hand, it is related to national economic policies, successful technical innovations, and security policies. Moreover, the differences are due to the different cultures and habits in countries and societies involved. In Germany, for example, the right to data protection is widely developed and routed in basic law, requiring the government to secure its citizens' rights (Art. 2 I and Art. 1 I of the Basic Law of the Federal Republic of Germany). In the European Union—a body of common cultural habits with Germany as a major stakeholder—the main goal of cyberspace policy is to strengthen systematic resilience and the ability to recuperate from attacks and fraud.

Americans, on the other hand, are accustomed to the view that it is not the government's responsibility to secure their personal rights. U.S. law derives from a fundamental right to privacy which includes the Fourth Amendment to the United States Constitution (with the freedom from unwarranted search or seizure), the First Amendment right to free assembly, as well as the 14th Amendment due process right, generally described as "the right to be left alone." U.S. cyber security policy itself is, contrary to the European view, driven by the military logic of deterrence.

Cross-Border Public–Private–Partnerships

In America's and Germany's service-based economies, most economic transactions carried out by essential economic institutions and critical infrastructures such as energy companies, the healthcare, the banking and the transportation sector, depend on sustainable networks. In order to directly ensure the availability of infrastructure and transition via partner companies or agencies, most countries have to rely on collaboration with their national internet service providers (Bendiek 2014), which themselves have to rely on unlimited access to global networks for establishing secure data transfer.

Another option to ensure the availability of infrastructure would be to ensure the management of the Domain Name System Root. Root servers are driven by different institutions in several countries, but they are coordinated by the Internet Corporation for Assigned Names and Numbers (ICANN). This is why some countries believe that the U.S., home of the ICANN, manages the DNS Root in a way that benefits the U.S. system more than others. In this view, the alternative would be to establish a more regional system for DNS Root management.

Cross-border and cross-sectoral cooperation between governments, private organizations, and companies is the logical - and unavoidable - consequence of these considarations.

Data Sovereignty Within Local Markets

To cope with this challenge, Germany and France were considering a so-called "Schengen Cloud" system for data within the European Union (Bendiek 2014) in 2014. Such an EU cybersecurity policy would be closely linked to national and international regulatory processes and would therefore need to be formulated and implemented on a global multi-level and multi-stakeholder structure. In such a European system, as much online data as possible would be kept in Europe. However, it is quite uncertain whether this system would indeed limit surveillance and not just prevent EU corporations from access to economic markets. The idea, proposed by Deutsche Telekom AG, might function within the scope of the Schengen Agreement, which stands for free movement of people and goods across participating EU Member States, as long as the Schengen system remains operable.

The Iranian government, to pick another example, reportedly allocated US$500 million in its 2010–2011 annual budgets for the purpose of combating what it termed a "Soft War" being waged against the regime by its perceived enemies via media and online activities (Freedom House 2015). The development of a national infrastructure would give the Iranian authorities full control over access to the internet as well as the ability to monitor all content that is transferred within its national infrastructure. China is famous for successfully employing a similar tactic by using its own services and software developed for China's citizens. The Chinese block foreign services and monitors all data transferred via the Chinese grid.

But also several possible technical measures created to secure the real transfer of data, as well as the infrastructure, are on the table. One possibility would be to implement Domain Name System SECurity (DNSSEC). This would enable users to determine the ability to decipher if the data has been changed during transport. Another possible option would be to force ISPs to establish a process or framework for border-crossing data transfer. The Border Gateway Protocols (BGP) includes technical operations and protocols ensuring that routes cannot be redirected or disrupted in any way. However, hijacking of BGP has already taken place on many occasions, increasing the desire of many countries to secure their national grid structure. For them it is essential to know the paths their data take, and where and how it is stored (Hathaway 2014).

Sovereign markets face yet another challenge: The regulation of governmental access to extraterritorial commercial data. Many companies collecting huge amounts of customer data such as Google, Yahoo or Amazon are, for example, headquartered in the U.S. The resulting asymmetric access of U.S. law enforcement agencies to foreigners' information is seen partly as an infringement of nation states' sovereignty and partly as a violation of domestic laws. The answer to this should be a multilateral or at least a bilateral agreement on how law enforcement requests for extraterritorial private sector data should be treated under the governing law.

Existing Treaties as Regulatory Framework

Sovereign countries have been arguing over protective measures related to data transfer ever since data transfer has played a role in international business and communication amongst governments. The electrical revolution in the nineteenth century appeared with telegraph wires crossing national borders. Where lines crossed national borders, messages had to be stopped and translated into the particular system of the next jurisdiction. In response, European states created a framework to standardize telegraph equipment, adopt uniform operating instructions, and lay down common international tariff and accounting rules. Along these lines, the original International Telegraph Convention in 1934 established first measures to stop messages that may would have had an impact on the safety of the nation state or would have been contrary to the laws of that State (Hathaway 2014).

Roughly the same is happening today with data crossing borders via the internet. The 2012 International Telecommunication Regulations (ITR), in effect today, does not contain explicit provisions for securing traffic and supporting IT infrastructure. It does, however, include an exemption clause to avoid "technical harm" (Art. 9.1 b) (ITR). This article, inaugurated in the 1980s, may lead to the impression that national governments are allowed to interfere for any national security reasons. It was added in response to the Morris worm in the 1980s, one of the first pieces of malware.

The treaty is governed by the International Telecommunication Union (ITU), a specialized agency of the United Nations (UN) that is responsible for issues that concern information and communication technologies. ITU membership is open to governments, which may join the Union as member states, and private organizations such as carriers, equipment manufacturers, funding bodies, research and development organizations and international and regional telecommunication organizations, which may join the ITU as non-voting sector members.

There are suggestions to add regulations in the treaty to include explicitly security-related cases. However, there is still no globally accepted definition of cybersecurity. This obviously obstructs protection efforts. As a consequence, incidents are usually treated under national law within existing laws and regulations. This will last until the international community will do have in place international standards concerning relevant attacks, cyber and information warfare, cyber security, and other related terms. Therefore, ITU's Telecommunication Standardization Sector started to publish standards for cybersecurity. In addition, ITU assists developing countries with special guidance. In its Global Security Agenda May 2007, the ITU supports cooperation to promote cybersecurity and enhance confidence and security in the information society on an international level. This is part of its mandate to lead the coordination of international efforts in "building confidence and security in the use of ICTs".

Tallinn 2.0 Project

As a response to growing threats to sovereign countries and critical infrastructures, a group of well-known law experts from NATO states met in 2013, upon invitation of NATO's Cooperative Cyber Defense Centre of Excellence. In February 2016, the legal expert members of the so-called Tallinn 2.0 project (Schmitt 2013) on international law stated that there are already multiple international law regimes appling to military cyber operations. The manual itself declares that, in principle, the provisions of the Charter of the United Nations are applicable to cyberattacks (Tallinn Manual 2013). They predicted that there will still be much debate on the extraterritorial reach of the treaties and laws, particularly on its precise application to such matters as monitoring communications or collecting metadata in the near future.

Acceptance of Uncertainty

The safeguarding tendencies of sovereign countries are only reasonable given the effects cyberspace has on each sovereign country, including its security, and economy. Nonetheless, it should be taken into account that they lead to a fragmentation of the internet which was originally built with the World Wide Web as an open space communication platform. The internet developed with an open access approach for everyone. Although originally was not intended for economic use, it soon became the world's busiest marketplace with states as stakeholders fighting for superior strategic access and power.

The internet functions as any other market does, except that it originally had no geographical or state-related boundaries. Even though there is no internet security that can be defended, the digital society lives in an insecure environment that can only be protected by appropriate technical measures and reactions of governing bodies. This requires a physical infrastructure, e.g. encryption technologies. Encryption as the process of encoding messages or information in such a way that only authorized parties can read it does not of itself prevent interception but it can help to secure communication by denying the content to the interceptor. In an encryption scheme, the intended communication information or message, referred to as plaintext, is encrypted using an encryption algorithm, generating ciphertext that can only be read if decrypted.

On both national and international levels, existing treaties and laws serve as an appropriate measure for securing the internet. Another suggestion is to rely on "due diligence" in cyberspace which is based on the international legal standard of "due diligence" (Bendiek 2016). It requires every state to do everything possible to prevent actions originating in its own territory. Other mandatory measures include minimum standards with regard to prevention, resilience, and international collaboration.

The internet's scope is the globe rather than geographically defined areas between landmarks. Therefore, a new and highly sophisticated monitoring is needed to appropriately govern the internet. The multi-stakeholder approach of the internet is at this point one reason why it is so difficult to maintain and to manage between governments, companies, and citizens.

Originally, the internet was a publicly financed project of the U.S. government. The privatization of the internet and its governance started in the 1990s and is still an ongoing process. Thus, leading governments should consider a more intense cooperation with each other, with the private sector, and other concerned parties. The fractious and inconclusive debate between stakeholders over surveillance practices in the transatlantic partnership, for example, has already weakened running systems and is still threatening resilient governance of IT infrastructures.

3 Economy

Online content, applications, and services are rapidly permeating all segments of commerce and society. They are affecting and disrupting traditional industries in many ways. Economic factors are thus the second main focus area of modern states in the distribution battle over the internet and the internet economy. By the end of 2015, 3.2 billion people were online (Facebook 2015). The remaining 4.1 billion, out of 7.3 billion comprising the world's population in July 2015 (United Nations 2015), were not yet able to connect to the internet. In general, the developed world is largely online these days, but the developing world is still lagging behind.

Population Prospects and Growing Markets
In the foreseeable future, the highest demand and market growth potential for connectivity and internet penetration will most likely come from Asia and Africa (United Nations 2015), bringing potential power and influence to their populations. However, a prerequisite for their internet access is the availability of the underlying infrastructure and technology that can deliver affordable broadband internet services to citizens in these regions. Their governments and companies are already laying the foundations necessary for providing universal access to their citizens, while simultaneously linking access to their economic sustainability and security agendas.

At this point, it is worth mentioning that China already has the largest number of internet users living in one country and the fastest growing number of internet users in the world. Secondly, most of the new IT software or computer hardware is currently manufactured in China's industry sector. This gives Chinese officials first-class access opportunities to data transfer.

Companies' Behavior on the Grid
Advancing internet connectivity requires facilitating network and broadband infrastructure expansion. These investments can be costly–and some countries may not have the resources to deliver high-quality, low-cost infrastructure to remote rural areas with smaller populations. In the last century, when landline telephone systems were more common, revenue was incurred through an inter-carrier international settlement system that negotiated a price per call based on origination and termination (Hathaway 2014). This collection system helped pay for improvements aimed at reaching more and more citizens. However, in today's internet protocol (IP) environment, ISPs either pay fees based on capacity or use settlement-free peering, thus bypassing the payment scheme previously imposed by inter-carrier international agreements.

Content providers that offer their services via the networks of infrastructure carriers' using an over-the-top (OTT) model pose further challenges to the pricing model. In Germany, OTT content and service providers include myvideo.tv, Google docs, Facebook, Dropbox, AppleTV, WhatsApp, Netflix and LinkedIn. OTT refers to content, services or applications that provide services to the end user over the open internet. The OTT service provider is typically distinct from the operator (ISP) of the underlying network and these service providers consume bandwidth through their delivery of volumes of information to users–usually for free. Sometimes this leads to poor quality of the infrastructure operators' own telecommunication services, given the fact that they are using more than their fair share of bandwidth. The net-operating companies are thus forced to invest in infrastructure to secure their on service quality. Governments should weigh up whether they recognize the need to implement new pricing models.

Reactions of National Leaders

As previously mentioned, both national and corporate leaders are threatened by this highly complex scenario. In this context, governments aim to increase their country's level of independence and to curb the negative effects of interdependence (Krastev 2014). The entities that control the flow of information can garner economic and political leverage in a country or a region. The perceived or very real inequality of who should be able to monetize access to the internet on the one hand, and who may benefit from that access on the other remains part of the ongoing controversy. Therefore, countries are seeking mechanisms to couple market access with cost-recoverable investments to pay for the modernization and expansion of the infrastructure that today's digital society is demanding.

In order to assert power over ISPs and OTT providers, some leaders are looking towards a regulatory environment and international treaty venues such as those convened by the ITU. Furthermore, the market liberalization of the past two decades may give way to the revivification of previously fully state-run telecommunications companies such as Deutsche Telekom, AT&T, and Vodafone, which, acting as ISPs, would be the wireway for citizens to reach the internet. This would give nations more control over private or semi-private providers, allowing them to channel the profits into their economy.

Depending on the perspective one takes, this could also be seen as a barrier to market access. For example, the German government stopped using Verizon Communications Inc. services as a service provider to German agencies by the end of 2015. The service was transferred to Deutsche Telekom AG (Holland 2014). This change was based on concerns about network security after the Snowden disclosures, due to the fact that Verizon is headquartered in the United States and acts as a core partner to the National Security Agency (NSA).

Cross-Border Data Transactions

Another important aspect of the economic side of the internet involves cross-border data transfer. The Transatlantic Trade and Investment Partnership, for example, seeks to increase economic growth for all signing partners. Many of the criticisms leveled at the Anti-Counterfeiting Trade Agreement, which lasted until 2012, reappeared (Krempl 2013). It is feared that internet freedoms are increasingly being subordinated to the logic of market commercialization.

All partners in these agreements will have to enable the free flow of data across borders if they wish to facilitate commerce. The European negotiators were already urged to prevent the undermining of EU data protection laws (Bendiek 2014). In parallel to these negotiations, process reforms and data protection in the EU have been ongoing since 2012. The old data protection law, dating back to 1995, prohibits the transfer of personal data from EU member states to countries that do not have privacy protection regulation standards comparable to those of the EU. The new data protection regulation shall apply from 25 May 2018. The directive entered into force on 5 May 2016 and EU member states have to transpose it into their national law by 6 May 2018. The new regulation will make sure that citizens' data is protected throughout the world, and not just within the EU. EU data protection standards will have to apply independently of the location in which the data relating to EU individuals is processed.

Other countries are seeking different ways to protect their data, declaring that there needs to be data sovereignty for national security purposes. Two very relevant cases come from the UK and U.S., both of which have adopted a protectionist stance in their policymaking. This controversy is particularly challenging in an era in which data is stored in multiple centers and locations to enable citizens' access on demand and to facilitate cross-border cooperation. Moreover, it raises fundamental legal and political questions, especially when it comes to international collaboration.

Governing Laws and Treaties in the Non-military Sector
The first question concerns the law applicable to private and criminal interactions. This law could either be related to the citizenship of the data creator or to the location in which the data is stored. The jurisdiction in question could change any time that data is multiplied, copied, transferred and stored again. The law changes when the transfer route crosses borders of sovereign countries.

Some countries may therefore wish to impose a law to inspect all transferred data, while others may demand that organizations use preferred service providers and store their data locally. Data will thus fall under local law, as already seen in Iran or in China. It is clear that such a confusing situation requires a multinational modernization of a rights-based framework for privacy and security policy. A modernized governing legal system should aim at increasing the clarity and legitimacy in the laws that apply. Secondly, it should implement an efficient procedure for the transparent implementation and enforcement of laws (The German Marshall Fund of the United States 2015).

In response to the massive increase in cybercrime activities, the U.S., Canada, South Africa, Germany, and Japan signed the Budapest Convention (Council of Europe 2001). In March 2016, nearly 50 countries ratified the treaty. The convention is the first international treaty to bring together national criminal laws and prosecution of internet-related crimes. The convention came into force in 2004, covering a wide range of criminal offenses in an attempt to compensate for gaps resulting from differing national criminal laws. It sets standards and defines criteria for ascertaining whether crimes have been committed and lists measures with which to react if laws are breached.

Extended Grid Structure

Apart from that, huge efforts to promote the development of Internet Exchange Point (IXP) facilities to enable the fast transfer of data through IP interconnections have been carried out on a large scale. As countries strive to connect citizens in rural areas, they will need multiple IXPs to ensure low-latency delivery while striving to ensure end-to-end quality of service. This will force carriers to take measures to further the security, safety, and resilience of their own infrastructure. The involved companies will have enormous power over data transfer and the content itself as they are natural partners of the governing bodies. As contractors to citizens, they likewise possess great power over citizens' private lives and should therefore be or become close allies of governments.

4   Society

Internet access is of tremendous importance for the free market, for democratic decision making, for citizens' daily lives and thus for the future of democratic order. Therefore, the conflicts between several states as mentioned before on the treatment of personal data have blasting power. Due to the fact that the right to privacy has a direct impact on citizens' life, it is necessary to consider that privacy is not only related to international politics, but also to domestic politics. Privacy rights are directly related to the relationship between national governments and their citizens. Therefore international negotiations about privacy rights should be conducted and be managed according to the social customs of each society involved.

The NSA's spying activities have not only affected the transatlantic partnership but also the relationship between the German government and its citizens. Surprisingly, the actions of the NSA haven't met with serious protests from governments all over Europe nor have they provoked sharp answers from Berlin. They did, however, reveal the close collaboration between the German intelligence agencies and the NSA. This was perceived as a breach of trust between the German government and its own citizens. The long-held myth that the World Wide Web is an unrestricted open space, with access for everyone, without restrictions, without control of or in relation to institutions and governments has proven to be an illusion. It first became a marketplace, then a place for free communication around the globe, and will soon turn into the site of strategic negotiations about who will be shaping public opinion, and about geostrategic and economic positioning.

The speed and the immense scope of the transformation of our communication habits via the internet and wireless communication have initiated all forms of reactions from politicians and citizens (Bargh and McKenna 2004).

For a better understanding, it is then necessary to place the internet in the context of the transformation of the overall social structure, as well as in relation to the cultural characteristics of this social structure. The new social structure established by the communicative capabilities of the internet is characterized by the rise of a new culture of autonomy.

History of the Grid

The internet was initially financed by the U.S. Pentagon for the benefit of scientists, engineers and students and originally had no direct military application. Nowadays, the rising number of military actions in cyberspace speaks quite a different language. Since the internet's expansion in the 1990s started with the technological discovery of the World Wide Web at CERN, the European Organization for Nuclear Research in Switzerland, the web as an information space has been running under the principle of open source. A next major step was the institutional change in the management of the internet; it was handed over from the U.S. authorities to the loose control of the millions that make up the global internet community. In this way, its governance was privatized. As the internet started to be privatized, it soon allowed for commercial purposes as well as (open) cooperative uses, and became more and more a battlefield.

Societal Changes and New Communication Habits

These developments led to enormous changes in social structures, cultures, and social behavior around the globe. Networking started to become the prevalent organizational form among citizens. Social online behavior became increasingly individualistic. The culture of autonomy suddenly became the culture of the network society.

These social changes must be taken into account in relation to political decision-making. The transformation of communication and changes in social structures, culture and behavior also changed the citizen. This should be of particular interest to politicians, as some of these processes have a direct impact on political decision-making.

Power and resistance to power, as well as fundamental relationships within society, are constructed in the human mind, through the construction of meaning and the processing of information according to specific sets of values and interests. Ideological apparatuses and the mass media continue to be key tools of mediating communication and asserting power. But the rise of a new culture, the culture of autonomy, has found a major medium of mass self-communication and self-organization through the internet and mobile communication networks.

For example, the so-called 'Facebook refugees' chart their escape from Syria on cell phones in a fully self-organized manner (CNN 2015). They arrive with the help of smugglers at the Greek border, and already know all the necessary information to ask of the local officials. Others gather in Turkey or in Libya to cross the state borders in huge groups arranging their gatherings via Facebook. Surprisingly, in 2015, social networks as a tool for self-organization in a culture of autonomy were supporting one of the largest refugee streams seen in a long time.

Civil Rights, Labor and the Internet

Nowadays, it is hardly an option for Western governments or citizens not to use the internet. As a result, it is important for citizens to determine if the internet should be a civil right or a privilege. A few years ago, the Freedom Online Coalition was formed as a partnership of 29 governments, aiming to advance internet freedom. Coalition members work together closely to coordinate their diplomatic efforts and engage with civil society, and the private sector to promote free expression, association, assembly, and online privacy worldwide.

The 2014 NetMundial meeting of the group drafted an agreement on human rights aiming to underpin internet governance principles (NETmundial EMC 2014). The focus of the Global Multistakeholder Meeting on the Future of Internet Governance is to negotiate internet governance principles and to draft a roadmap for the future evolution of the internet governance ecosystem. Following the UN Human Rights Council's 2012 decision, they declared that people's offline rights should also be guaranteed online. This means that existing international human rights treaties and legal obligations should be applicable to online activities, too. These rights include freedom of speech, freedom of association, privacy rights, as well as freedom of information and access to information.

The internet has also shifted power and perceptions with regard to intellectual property rights. Many citizens have lost their income from intellectual property rights or had to refocus their career paths, traditional ways to negotiate and to sell goods have faced fundamental changes, with tremendous consequences, particularly in the music industry, in technical industries like the camera industry, and the printing press (Keen 2015). On the other hand, major companies such as Google, Yahoo, and Facebook are on the rise, creating many new jobs.

Violation of Rights and Liberties

Governments are believed to be infringing on citizens' right to privacy. As a result, there is widespread disagreement as to the extent to which governments should be able to access private data. As mentioned before, the ongoing discussion on cyberspace regulations is shaped by conflicting views on how to achieve a fair balance of interests. These differences over the right balance of interests are caused by differing geostrategic positions of the nations involved. The questions concern the fields of organized crime, terrorism, as well as the issue of how long private data may be stored and whether it can be used for purposes other than initially intended. To address these concerns, former U.S. President Obama, for example, decreased the scope of the NSA's collection activities.

Many other Western countries are reviewing the fields of operation of their surveillance agencies and related organizations under their guidance, an example being two new laws passed in Germany regarding the Federal Intelligence Service and the Federal Office for the Protection of the Constitution. On the other hand, countries have already put laws on data collection into force. Germany has implemented a law on data retention. On 2 March 2010, the German Federal Constitutional Court rejected the legislation requiring data retention of electronic communications traffic for a period of six months. The court judges agreed that data storage was not secure enough and that it was not clear what it would be used for. The court ruled that such retention represents an especially grave intrusion into citizens' privacy (Bundesverfassungsgericht 2010).

Governments are keen on collecting data and can even force private companies to collect data for them. The Federal Bureau of Investigation, for example, tried to compel Apple to mine data from an iPhone used by one of the shooters in December's 2015 terrorist attacks in San Bernardino/United States (Nakashima 2016). Many countries are passing rules in these fields. If companies choose to sell their software programs to government agencies or administration they can thus be forced to deliver their source codes.

On a parallel line of thought, another question emerges: Should those being governed should have the right to know what their governors are doing with their collected data. Not long ago, the European Court of Justice ruled in favor of a Spanish citizen and against Google (European Court of Justice 2014), thereby implementing the new right of European citizens to be "forgotten" or the right to have information of concern deleted from the internet.

To implement sufficient laws on the use of the internet it should soon be made clear that solutions must be aligned with principles of human rights, should be responsive to the complex political economy of surveillance policy, and be premised on common interests and values (The German Marshall Fund of the United States 2015).

5   Conclusions

Germany's cybersecurity efforts reached a new level in October 2016 with the launch of the White Paper 2016. Nevertheless, no country is cyber-ready yet and cybercrime has become a business which exceeds a trillion dollars per year in online fraud or theft, affecting millions of people around the world.

Four central problems for democratic governance of the internet have to be recognized while implementing adequate cyber strategies on a national and international level:

(a) **Fading boundaries between internal and external policies** have produced new approaches of intergovernmental cooperation, multi-stakeholder regimes, and multilateral bodies. Governments, international organizations and working groups should take this as a starting point of a long-term negotiation process on this topic. Developing countries and areas should be invited to be part of these negotiation processes from the beginning.

With the increasing expansion of information and communication technologies and the growing chance for real-time boundless exchange, cyberspace as an operational domain for companies, citizens, and governments, is a complex transnational issue that requires global and intersectional collaboration for ensuring a safe internet environment.

Breaches of trust caused by surveillance scandals in international partnerships should be seen as long-term challenges for bilateral and international cooperation. The same occurs when the internet is used for purposes of warfare, intellectual property rights theft, online fraud, attacks on critical infrastructure, and other private companies.

(b) **Protectionism by governments** is reasonable due to the distribution battle over the internets' possibilities to gain power over sovereign countries, their citizens' data, and over market shares in the internet. Since the internet is a market like any other market, except that it has no original geographic or state-related boundaries', and emerged as an open information space it became necessary to implement an internet governance.

Moreover, the internet as a technology has a material culture. Technical, regulative, and legals means are therefore needed to provide a secure environment for users in the World Wide Web. These new regulations should be aligned with principles of human rights, they should respect the security needs of sovereign countries, take into account the complex political economy of surveillance policy, and be based on common interests and values. Other measures required to create a secure and well-governed space should include minimum standards with regard to prevention, resilience, and international collaboration.

(c) **The privatization of internet governance** started in the 1990s when the U.S. authorities handed it over to the loose control of private companies running the systems, collecting data, manufacturing the devices, and to the millions that today make up the daily growing global internet community.

Such developments led to changes in social structures, cultures, and social behavior around the globe. Networking started to become the prevalent organizational form among citizens. Social online behavior became increasingly individualistic. The culture of autonomy suddenly became the culture of the network society, which should be of particular interest to politicians, as some of these processes have a direct impact on political decision-making.

(d) **New forms of cooperation and participation** with relation to the internet occurred on different levels. These must be implemented between sovereign states to solve problems emerging with regard to internet governance. As seen in different negotiation processes on a multi-stakeholder and on a transnational level, several states, especially the third world countries, are fearing to be left out when not represented in governing or policy-making bodies.

Due to the fact that governments must rely on private companies to directly secure internet infrastructure and transition of data, they should involve all stakeholders concerned when developing a new internet government system.

Another highly important aspect in this context are human and civil rights. The right to be connected to the internet, respect for intellectual property rights, protection of personal data, and citizens' protection from the government must be addressed by new treaties and regulations on a national and transnational level. Other important rights include the freedom of speech, the freedom of association, privacy rights, and the freedom of information and access to information. Governments should be wary of losing contact with their citizens, a danger which should also be considered a long-term challenge for democratic development.

References

Bargh J, McKenna K (2004) The Internet and social life. Annu Rev Psychol 2004:55

Bendiek A (2014) Tests of partnership transatlantic cooperation in cyber security, Internet governance and data protection. Accessed 1 Apr 2016

Bendiek A (2016) Sorgfaltsverantwortung im Cyberraum. Berlin

Bitkom (2015) Digitale Angriffe auf jedes zweite Unternehmen. http://www.bitkom.org. Accessed 1 Apr 2016

Bundesregierung (2016) Weißbuch zur Sicherheitspolitik und zur Zukunft der Bundeswehr. Berlin

Bundesverfassungsgericht (2010) BVerfG, Urteil des Ersten Senats vom 02. März 2010—1 BvR 256/08—Rn. (1–345). Accessed 1 Apr 2016

Council of Europe (2001) Convention on cybercrime. Accessed 1 Apr 2016

Currier C, Moltke H (2016) Spies in the sky Israeli drone feeds hacked By British and American Intelligence. https://theintercept.com. Accessed 1 Apr 2016

European Court of Justice (2014) Google Spain and Google (ECJ C-131/12). http://curia.europa.eu/. Accessed 1 Apr 2016

Facebook (2015) State of Connectivity 2015. A report on global internet access. Accessed 1 Apr 2016

Freedom House (2015) Freedom on the Internet 2015 Iran. http://www.freedomhouse.org. Accessed 1 Apr 2016

Gabel D, Schuba M (2015) Germany rolls out IT Security Act. White & Case Technology Newsflash. http://www.whitecase.com. Accessed 1 Apr 2016

Gaycken S (2015) Why Germany's cybersecurity law isn't working. Council on Foreign Relations. 18 Aug 2015

Hammond P (2016) Chancellor speech: launching the National Cyber Security Strategy. https://www.gov.uk/government/speeches/chancellor-speech-launching-the-national-cyber-security-strategy. Accessed 2 Nov 2016

Hathaway M (2014) Connected choices: how the internet is challenging sovereign decisions. Am Foreign Policy Interests 36(5):300–313

Hathaway M, Demchack C, Kerben J, McArdle J, Spidalieri F (2015) Cyber readiness index 2.0. A plan for cyber readiness: a baseline and an index. Potomac Institute for Policy Studies. http://belfercenter.ksg.harvard.edu. Accessed 1 Apr 2016

HM Government (2010) A strong Britain in an age of uncertainty: The National Security Strategy. https://www.gov.uk. Accessed 1 Apr 2016

Holland M (2014) NSA-Skandal: Auch Bundestag beendet Kooperation mit Verizon. http://www.heise.de. Accessed 1 Apr 2016

Keen A (2015) The Internet is not the answer 2015. Atlantic Monthly Press, New York

Krastev I (2014) The interdepended World in crisis. In: Frank J, Matyas W (eds) Strategie und Sicherheit 2014. Böhlau, Wien Köln Weimar

Krempl S (2013) Transatlantisches Freihandelsabkommen: "Schlimmer als ACTA". http://www.heise.de. Accessed 1 Apr 2016

Nakashima E (2016) Apple vows to resist FBI demand to crack iPhone linked to San Bernardino attacks. https://www.washingtonpost.com. Accessed 1 Apr 2016

NETmundial EMC (2014) Multistakeholder statement. http://netmundial.br. Accessed 1 Apr 2016

Russian Federation (2011) Conceptual views regarding the activities of the armed forces of the Russian Federation in information space. http://www.aofs.org. Accessed 1 Apr 2016

Schmitt M (ed) (2013) Tallinn manual on the International Law applicable to Cyber Warfare. Cambridge University Press, Cambridge

The German Marshall Fund of the United States (2015) Transatlantic digital dialogue: rebuilding trust through cooperative reform. http://www.gmfus.org. Accessed 1 Apr 2016

United Nations (2015) World population prospects, key findings and advance tables. http://esa.un.org. Accessed 1 Apr 2016

Watson I (2015) 'Facebook refugees' chart escape from Syria on cell phones. http://edition.cnn.com. Accessed 1 Apr 2016

Exponential Technology Versus Linear Humanity: Designing A Sustainable Future

Gerd Leonhard and Carl-August Graf von Kospoth

The concept of the sustainable use of technology does not come from a luddite point of view. It is not at all about halting progress in its tracks or 'going offline' but about making technology more human, rather than making humans function more like technology.

We, both as a society and as individuals, urgently need to find ways to retain what makes us human, and to focus on which technologies actually make the world a better place for humans to live in. We are on a one-way track towards a world of exponential technological change in which only humans remain linear both now and hopefully in the future. This is the biggest challenge we face as we approach the pivot point of the exponential curve. Exponential growth, after all, is not just inexorable; it creates a widening gap between technology and ourselves. Unlike linear growth, where seven steps take you from one to seven, doubling takes you to 128, and in 30 steps you are at a billion!

This in turn means that the moment is just around the corner when machines will be more powerful than we are. And who's to say that 'thinking machines' will not be able to at least mimic human values in some way or another? So will humanity eventually be simulated by AGI (artificial general intelligence) that will keep us 'as pets'?

I don't have a dystopian view of the future but these are some of the questions we need to look at because humanity really is likely to change more in the next 20 years than in the previous 300 years.

G. Leonhard (✉)
The Futures Agency, Arlesheim, Switzerland
e-mail: Kerstin.von-Aretin@bmw-foundation.org

C.-A.G. von Kospoth
BMW Foundation Herbert Quandt, Munich, Germany
e-mail: Carl-August.Kospoth@bmw-foundation.org

© Springer International Publishing AG 2017
T. Osburg, C. Lohrmann (eds.), *Sustainability in a Digital World*, CSR, Sustainability, Ethics & Governance, DOI 10.1007/978-3-319-54603-2_6

Are we as humans prepared to deal with a future that will be shaped by the massive technologisation, digitisation, automation and robotisation of our society? We can't seem find the off-switch any longer and most of us certainly can't opt out because by abstaining, by limiting ourselves or by setting too many rules governing our use of technology, we would essentially put ourselves out of business. At the same time, in many companies we are already seeing a new kind of 'wired or fired'—ethos emerging—you either augment yourself with technology (from smart phones to augmented reality applications to, soon, brain-computer interfaces) or you won't be fit for the job.

The question really is how do we orchestrate these hyper-innovations so as to make them humanly sustainable? If humans remain linear but technology continues on its exponential path this will be the big challenge behind the sustainability question.

Never mind Moore's law and its inevitable 'end' as far as silicon chips go: we are in fact experiencing exponential growth in many other parts of our lives, too. Artificial Intelligence has really taken off in recent years, fuelled by the latest achievements in 'deep learning', neural networking and cognitive computing. Twenty years after IBM's Deep Blue beat Gary Kasparov at chess, Google's DeepMind recently beat the world champion Lee Se-dol at the Chinese game of Go, which is orders of magnitude more complex than chess. In fact it's said to be the hardest game in the world, with some 2.08×10^{170} possible moves (that's a number with 171 digits); more possible moves than there are atoms in the universe. And Google's DeepMind managed to do this without being programmed: it essentially taught itself the game. A great example when we consider how 'humanly sustainable' these technologies will or won't be in the future: self-learning computers are not very likely to tolerate human inefficiencies just because we are used to them.

By 2025, we will probably face the so-called singularity, meaning that we will be able to build and use computers that are as powerful as the human brain. But following the principle of exponential growth (even without computer chips actually remaining on that trajectory), this means that 18–24 months later the computer will be twice as powerful; in 4 years it will be four times as powerful, and by 2050 a single computer could be as powerful as all human brains combined.

The speed of our networks, the amount of data, the power of virtual reality headsets—everything is growing exponentially! The main reason we don't already have quantum computers is the issue of powering and cooling them: a single quantum computer would require more electricity to keep it at working temperature than a large city. Network technology is still stuck with relatively low-level connectivity like 3G and 4G or even LTE which won't be enough to support exponential technological leaps. Regardless of these current limitations, this is where we are heading. If everybody has his or her personal digital assistant living in the cloud, hooked up via neural transmitters directly to our brains, the question of sustainability suddenly becomes glaringly obvious: how will we remain human in such a super-charged world?

Of course, the idea of having an intelligent digital assistant isn't new. Hollywood introduced us to what a digital assistant could look like back in 2001 with the film

"Space Odyssey". From Microsoft's Bob to Apple's Siri, we have become more and more accustomed to relying on the handy little helpers which combine machine learning technologies from the fields of speech, natural language processing, and document analysis to provide a new way to interface our personal computing devices. And this is just for starters. As we progress towards wearable computing devices, these personal assistants will become more important because they provide an easier interface to access information on the go. But where will this technology go next? We may safely assume that it will shape the way we interact with computers on a day to day basis. And like Bertie Wooster, we may someday become totally reliant on our digital Jeeves, our mother's little helpers to get us through our busy digital days.

Technology is radically changing the way we see the world. Already, Facebook's CEO, Mark Zuckerberg is prophesising that virtual reality will become a very big part of everybody's life in the future. But as we spend more and more time immersed in a world of digital information and media, will our links to reality on this side of the computer screen or VR headset become weaker and weaker? Will we in fact still be able to discern between the "real real" and the "virtual real"? Some researchers worry that too much VR will have direct effect on our minds and bodies. We know that the experience can cause nausea, eyestrain and headaches. Jeremy Bailenson, a professor at Stanford University, says his 15 years of research consistently have shown virtual reality can change how a user thinks and behaves. Bailenson was one of the first, in 2007, to describe the "Proteus effect", a phenomenon in which the behaviour of an individual, within online virtual worlds, is changed by the visual characteristics of their avatar. This in turn can lead to a decrease in self-awareness and self-evaluation, he maintains, a phenomenon he calls "deindividuation". Virtual experiences, he believes, can change the cognitive structure of our brains.

According to Bailenson, there is a growing portion of our population that views face-to-face interaction as the exception. They would rather be viewing content on Facebook than talking to people like you or me. He is a strong advocate of limiting the time we, and especially our kids, spend in virtual worlds.

Thankfully, today there are still some limiting factors, but in a few years we will find that these machines are not only more powerful than we are in terms of raw computing power; they will also be able to learn and to 'think' (albeit still lacking social or emotional intelligence). If a computer is capable of learning, it will be able to learn the underlying emotional patterns that influence the human thought process. These machines will still not 'be' in the human sense—but they will deliver really powerful simulations of being.

Of course, computers don't really think in the same way that we do, but then again, they may not need to, either—in fact, we still haven't really discovered how human thought processes actually work. All computers do is to simulate what we think of as thinking. They are very good at processing information or handling complex topics, but human thought is embodied, i.e. correlated with our bodily existence, and it involves millions of interdependent flows that we still do not fully understand.

For instance, we still only have rudimentary knowledge of how exactly neurons distribute signals and create synapses linking seemingly unrelated pieces of information. Even on an exponential curve, this will take many years until we can unravel the mysteries of human intelligence and human consciousness. "I think, therefore I am" as Descartes famously said—but how will this apply to machines?

The U.S. government recently launched the Human Brain Project, a multi-billion dollar research project that hopes to reach major break-throughs in 'machine thinking' in the next 10 years. If they are successful, we might be able to reverse-engineer human brains. The model is the Human Genome Project, which unravelled the secret of human genomics in a similarly short time period.

But if computers are really capable of learning, they will be able to understand, or at least to mimic the emotional patterns that shape human behaviour—all they need is enough data and connectivity. Their ability to 'understand' emotions will be based on brute force computing and deep learning.

What if a computer gets to look at, say, 300 million retirement records, pension accounts or social security payments and also gets to monitor all the phone calls from people calling in to complain about some problem with their benefits. It will be able to analyse voice tones and facial expressions and deduct recognisable patterns. After the computer has handled a few billions of these transactions it will have figured out if there's a rule behind all of it. It's not humanly possible to do it that way, but it may also be unnecessary to teach the computer these rules because it will be able to figure them out by itself thanks to Deep Learning.

At that point we humans will no longer able to judge whether the computer is wrong or right because we will not be able to understand how the computer got there in the first place; that's just too many facts for human computation.

So how will humans remain in control? Is the growth of exponential technology humanly sustainable? That is the underlying issue with AI and deep learning which is all the craze in Silicon Valley these days. Computer scientists are excited about the idea of no longer needing to program a computer but instead let it teach itself, for instance, to beat the human champion at the game of Go, just by observing and analysing every single piece of information there is about Go and playing the game a few million times until it is unbeatable, at least by any human Go player.

At that point we will become unable to gauge whether the computer is doing its job because we simply cannot grasp that kind of complexity. Some people have suggested that we need to build some kind of ethics into our computers. As far back as 1942 the science-fiction author Isaac Asimov devised his "Thee Laws of Robotics", which stipulated that a robot must first never harm a human being, that it must second obey all orders given it by a human being and that it must third protect its own existence as long as this does not conflict with the first and second laws.

Unfortunately, these laws no longer suffice. If a computer is unable to harm a human, what happens if, let's say, a self-driving car is forced to harm a human in order to protect another human, maybe its driver as opposed to a pedestrian. Absolute laws like Asimov's will cease to work here. A medical robot in your home might be compelled to force you to take your medicine even if you, the patient, don't want to—it's good for you, dear! Or what if a robot needs to stop you from doing something stupid like driving while drunk? Autonomous, learning machines are far beyond Asimov's more or less mindless robots and their simple set of laws.

We have already reached the point where our position of control over machines is no longer sustainable in the yes/no way it used to be. So what we need is a new set of Laws and Ethics of Robotics and of Artificial Intelligence that will allow us to control our machines without necessarily understanding their computations which of course will be tough. In my view, foundations such as the BMW Foundation, Non-Profits and NGOs now have a unique opportunity to add significant value by catalysing and bundling these conversations on the highest political and societal levels.

The march of progress in technology often proceeds from magic to manic to toxic. At first it works, it's fantastic—in other words, it's magic! Then we start to get obsessed with it because it works so well that we can no longer do without; it becomes habit-forming and addictive. And finally, it reaches the toxic stage where it begins to poison our relationships with other humans, and with ourselves. A truly sustainable way of using computers would be to take yet another step, namely by devising rules and laws which can govern machines which are more powerful then ourselves.

What can we do, both collectively and individually, to tame the beast we have released upon the world? First we need to face the problem, something which most people are very reluctant to do. We need to make decisions individually and collectively to keep sustainable to ourselves. For instance we could choose not to keep our smartphones on the nightstand or switch them off not only during flights, but at certain times of day or during holidays. On average, people in the United States across all age groups check their phones 46 times per day, according to Deloitte. That's up from 33 looks per day in 2014. Is this really necessary? Is it healthy? The choice is ours but the addictive forces of technology are growing stronger every day.

Constant mental overload caused by a tidal wave of incoming signals and information could severely cripple our creativity, and reduce our overall quality of life. In fact, this kind of 'digital Viagra' could eventually render us 'mentally impotent', as well. The idea of 'living in the moment' could become a thing of the past—or at least become much harder to achieve. Is a life without any boredom, without empty spaces, actually a good thing?

Digital obesity is becoming a problem. We're getting fat digitally, so to speak, and the effects are similar. It's actually a lot like over-eating: we have too many things to eat, and if we try to eat too much we grow bloated. The same processes that have led to dramatic overweight in many Western societies are being used by corporations to induce us into wanting to eat more are also being used to get us to spend more and more time online.

Of course, generating this kind of craving for food or for digital experiences are both very powerful business models. Just because all of this information is becoming instantly and freely available, do we need to consume it at all times? Do we really need an app to tell us where in the store the music section is located, and do we really need to count our steps so that our fitness status can be updated on a social network? Unless we find a way to deal with this constant tsunami of possibilities, we may ultimately all become digitally obese, or even worse, become a part of the machine ourselves.

Exponential technological progress can have potentially very harmful—even if unintended—consequences including:

- Outsourcing 'thinking' to intelligent software, the 'smart cloud' and mobile digital assistants like Google Maps ("Look to the right: your destination is in front of you")
- Outsourcing of personal judgment to online peer platforms like Tripadvisor ("This restaurant is rated #1 in your current location"). Never mind that no local punter in their right mind would ever set foot in it, and that all the 5* reviews are from tourists that have just arrived and simply followed the inflight magazine's recommendations. . .
- Appification and substitution of human conversations, interactions and decisions through apps like PeopleKeeper, which offers to monitor your anxieties when communicating with 'friends' and will then offer to delete the 'bad' connections from your personal network.

If we want to avoid the horror scenarios that would result from a thousandfold boost of what we are already facing today, we need to accept responsibility, both as individuals and as a society, for choosing which decisions we want to delegate to technology and just how much control we wish to maintain over our everyday routines.

Technology, of course, can be a huge boon to society. Far from being a zero-sum game, it can increase our cultural diversity and enable us to make better decisions; to in fact become better and more intelligent ourselves. But we also share a personal responsibility for keeping our own lives sustainable and ensuring that the ultimate outcome is not a system so complex that in the end it becomes completely non-transparent to us humans who, at least theoretically, are the masters of the system.

Governments, public officials and organisations need to put pressure on the companies behind this trend. Collectively, we need to understand that technology is an extremely powerful business and that many of the things they create are highly addictive—like many other drugs but without a lot of the social stigmas attached to it.

We will need to regulate the drivers of digital addiction and discuss how to make these issues a lot more transparent. Technology should not be addictive by design. Many social networks employ behavioural scientists working on how to make the platform more addictive—that is not a good thing. This is like cigarette companies adding substances to tobacco that make smoking more addictive.

As a society, we have come to ban such corporate behaviour and we need to find similar rules as to how we keep people in balance. This is a societal question, not just a private one.

In the end we will need to agree on simple rules about how to keep technology sustainable. I call this "Digital Ethics"—and these must almost by definition shape the core of any truly sustainable technology. Digital human rights and digital ethics also need to be established as core rules for doing business in the technology market.

The list would be long and we need to come to some form of transnational agreement on them, but I am pretty sure they need to include rules like these: we should never allow humans to be governed by technology, specifically AI. Technology should never cut humans out of the loop just to assure speed or profit. We should not augment humans to achieve super-natural powers. We must not empower machines to upgrade and expand themselves without human consent. We should retain our right to disconnect and remain embodied as humans.

This right to disconnect is under threat today, and we need to push back. We also need to be able to take our time, and we must not punish people for tuning off and signing out. Humans, unlike machines, need time for contemplation and digestion. We are under increasing compulsion to be always online, and that will only get worse. For instance if you disconnect your car, your insurance company could cancel your policy because they are no longer able to track you and monitor your driving behaviour.

Perhaps the most important rule of all should be: we must not seek to completely eliminate accidents from happening, to suspend the laws of chance or extinguish serendipity, that quintessential human quality of finding meaning in apparently irrelevant or happenstance discoveries. As the old saying goes: the time you enjoy wasting is not wasted time. No computer in the world will ever be able to understand that.

As a BMW Foundation Herbert Quandt we have understood that there is considerable scope for action on the part of providers, users and society at large in adopting a differentiated approach to dealing with the risks and opportunities of digitalisation. In tackling the conflicting forces at work here, we will be offering various formats and programmes aimed at supporting managers in addressing these issues through concrete projects that seek to "benefit the organisation internally as well as society at large".

Humans in the Loop: The Clash of Concepts in Digital Sustainability in Smart Cities

Christiane Gebhardt

1 Introduction

Despite the fact that Green IT has increased efficiency and resource optimization and has also opened up new business options in the environmental sector, there is ample empirical evidence that substantial improvement on sustainability issues such as climate, deforestation or biodiversity relating to United Nations' sustainability reports (see, for example, UN 2016) has not taken place in either the better world, the smart city or the circular economy.

Why is that? We venture the thesis that digital ecosystems and sustainability have been misunderstood and misinterpreted in terms of underlying concepts and values, and in their respective requirements for governance and control. The paper sets out to discuss digital ecosystems in the context of the smart city to illustrate this argument, and elaborates on patterns that connect the digital and the sustainable worlds as well as on the differences that separate them. The debate culminates in the requirements for good governance of a de facto unmanageable complexity during the transformation of cities towards sustainability. We believe that cities exemplify the problem of governance, illustrate limits of control and shed light on the need for a new generation of controls that identify and mitigate the rifts, the fears and the unexpected in the process of integration and innovation in socio-economic systems.

C. Gebhardt (✉)
Malik Institute, Geltenwilenstrasse 18, 9001, Sankt Gallen, Switzerland
e-mail: christiane.gebhardt@mzsg.ch

© Springer International Publishing AG 2017
T. Osburg, C. Lohrmann (eds.), *Sustainability in a Digital World*, CSR,
Sustainability, Ethics & Governance, DOI 10.1007/978-3-319-54603-2_7

2 Integration of Concepts and Differences

The integration of two conflicting concepts such as sustainability and digitalization brought about a change of paradigm in our understanding of social systems. How to govern complexity that comes with system behavior became a major challenge (Umpleby 2009). In the tradition of natural sciences, sustainability holds the promise of self-organization, self-adaption and viability and is linked to the self-organized change of ecosystems striving for survival. Stripped back to the bare essentials, ecosystems are feedback systems with long and short impact times aiming at endurance. In a forest, growth is balanced and system logics underlying the development of the ecosystem are based on system behavior such as principles of checks and balances, trial and error, functional redundancies to manage risk related to endurance, optimal use of resources, flexibility and ability to self-correct. The same principles are adopted to describe and design cyber–physical systems.[1] Reinforcing and counter-intuitive loops that strive for homeostasis of the system hold the course in the absence of a central operational center (Vester 2001, 2012).

In environmental ecosystems, coping with trouble is part of the requirement and viability of a critical mass for resilience of the system is the path towards survival. Loss of critical mass and points of no return caused by external disturbances as well as the entry of new members—such as invasive species or even slowly emerging interdependencies—can damage, or cause rebound effects in the system. Extinction of subsystems leaves room for new development and eventually niches will be filled with new life. According to this line of reasoning, mutation, adaptation, retention—change or the emergence of a new system—is the basic principle of life (Maturana and Varela 1987; von Foerster 1982). In environmental ecosystems we find a close connection between innovativeness and sustainability running in a fully

---

[1]The principles of cyber physical systems are (a) Robustness and adaptability: a resilient cyber–physical production system must be able to withstand external influences or be capable of adapting to disruptions, (b) Self-regulation and self-recovery: a resilient cyber–physical production system must be able to regulate its production process and recover by itself to the ideal state following a disruption event, (c) Short response time: fast development and implementation of a suitable response is required in order to accommodate process disruptions and minimize the time in disturbed mode, (d) Intelligent components: every component in a resilient cyber–physical production system must possess a component data model, containing information about its manufacturing and assembly operations, (e) Autonomous decision-making: every component is able to exchange information with the manufacturing station in order to negotiate and make decisions autonomously, (f) Redundancy: redundancy is incorporated in the architecture of a resilient cyber–physical production system, either by including several manufacturing stations that are able to process similar manufacturing operations or by including flexible operation sequences in the process that can be alternated according to current needs, (f) Dynamic disruption database: a resilient cyber–physical production system possesses a knowledge database of disruption scenarios and possible countermeasures, and (g) Escalation scenario: an escalation scenario simulation that takes into account several disruption events is required in order to enrich the decision support system (Galaske and Anderl 2016).

decentralized mode and on the minimum principle; that is, on the key question on how little is needed for survival.

Popitz illustrated that social systems develop on similar terms and end up differently (Popitz 1992). His study of a group of passengers on a sea cruise fighting for benches with towels and other means demonstrated that small and closed social systems strive for further exclusion. Elites have a tendency to destroy diversity and develop structural disparities; and, eventually, the system stagnates (see Popitz 1992, on the sea cruise; Rifkin 1995, on stagnation). Facilitation of upward mobility, integration of new players and inclusion of discriminated groups are levers to support innovativeness and build up resilience. More importantly, values, sense-making schemes and belief systems within social systems are constitutional variables that need to be taken into consideration (Weick and Quinn 1999). Social reality is constructed (Berger and Luckmann 1966) such that hard facts and concepts are interpreted differently, and interpretations may change the behaviour of the system.

Consequently, the recent debate on the 'smartness' or 'happiness' of cities includes fuzzy concepts such as transparency, participation and well-being (Florida et al. 2013). The debate reflects that social systems do not run on functionality and fitness alone. Unlike our garden pond or the forest, social systems need more than openness, optimal interconnectivity and intelligent feedback loops for adaption and survival. For instance, a shared belief system is a strong trigger for coherence and supports the prevalence of concepts, and ambiguity triggers imagination and creativity. The cultural dimension adds to the complexity inherent in the respective system. The cognitive complexity culminates in a hyper-complexity when an organized system is capable of reflecting itself—see Fig. 1 (Göllinger 2012).

Despite all the differences there is a pattern that connects: in ecosystems and social systems complexity is not a problem but, rather, a condition sine qua non for the prolific development and viability of the system and must not be seen as a showstopper for digitalization. IT experts in system dynamics quantified complex

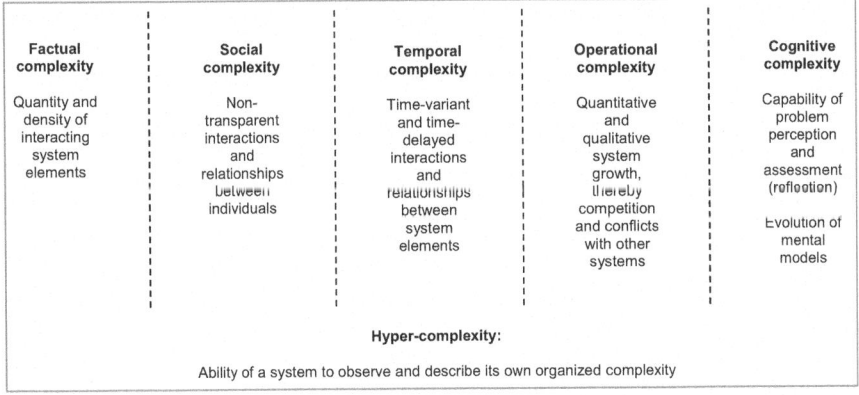

Factual complexity	Social complexity	Temporal complexity	Operational complexity	Cognitive complexity
Quantity and density of interacting system elements	Non-transparent interactions and relationships between individuals	Time-variant and time-delayed interactions and relationships between system elements	Quantitative and qualitative system growth, thereby competition and conflicts with other systems	Capability of problem perception and assessment (reflection) Evolution of mental models

Hyper-complexity:

Ability of a system to observe and describe its own organized complexity

Fig. 1 Dimensions of organized complexity. Source: Göllinger (2012)

social behaviour, and modelled and applied it to urban dynamics, as early as 1969 (Forrester 1969).

3 Digital Ecosystems

In innovation theory, an innovation ecosystem is defined as 'a network of interconnected organizations, connected to a firm or a platform that incorporates both production and use side participants, and also creates and adopts new value through innovation' (Autio and Thomas 2014: 205; see also Ritala et al. 2013; Adner and Kapoor 2010).

In digital ecosystems players are related via communication channels and business models: resource dependency, profitability and value proposition are the principal system logics. Digital ecosystems follow global business rules to gain a competitive edge of the cluster. As soon as the hierarchy-free, transparent and open discourse of IT networking is linked to business, the situation and the mode of functioning changes (Porter and Kramer 2011). Digital ecosystems catch and release skilled experts and workers, build and abandon infrastructure on a global scale, and so redefine space. They upgrade and transform regions and cities by providing an IT governance structure as a backbone for decision-making and navigation in energy, mobility, information and resource management. It is important to recognize that interdependencies in these digital ecosystems are by design and man-made: they are based on managerial decisions. In this way organized social systems evolve in which 'homo economicus' seeks profits, minimizes the risk of extinction, and enhances profitability, efficiency or growth. Managers employ both a business and a connectivity model based on a set of rules and regulations in which they believe. In digital ecosystems, socio-economic system logics do not necessarily lead to the same development as in sustainable development. As Googins noted, 'While business is solidly anchored in its self-interest, these more social, environmental and political issues are not set in its natural turf' (Googins 2013: 90). Unfortunately, transition and structural changes in the digital ecosystem may cause disruption in the social and environmental world. Loss of jobs, the abandonment and decline of cities and regions, and environmental problems, are the short- and long-term results of reinforced loops and non-intended consequences in the system. Equally, hot-spots, urban renewal and public services solutions materialize in the same city. As a general rule digital players are not accountable for social and economic problems concomitant with unequal development. Problem-solving in unrestricted situations tends to be assigned to the political level for the definition of an adequate framework and for intervention. Equally, urban activism and benevolent industrial engagement in cities play a role to bring about change (Sassen 2003). This brings us to the revised role of governance and complexity management in digital ecosystems.

4   Complexity and Limitation of Governance in Cities

Cities have often been perceived as living laboratories for transformations of the backbone systems addressing mobility, waste, food or energy and their effects on property, urban space and population and social peace: what it will be depends on a specific set of different factors working in a complex system of interrelating variables. In cities we find the intermingling of spheres as well as the spontaneous formation of viable, critical masses of new developments and concepts muddling through. The idea of analyzing cities using a systems approach and with an eye to the transformative dimension is not only intriguing, it is also highly relevant in times of increasing urbanization and new digital possibilities (Deakin and Leydesdorff 2014). At the dawn of the Sixth Kondratieff we analyzed data flows generated by human behavior with the aim of finding patterns of connectivity and identifying systemic risk, in order to find answers not only for understanding how fuzzy relationships function and systems produce results, but also to find models of new governance and second-order intervention to enable cities to organize themselves—in a desired, more sustainable direction (Gebhardt 2015). There is research dedicated to facilitate and accelerate intelligent intervention into system behavior to enable systems to repurpose themselves (Schwaninger 2006), mitigate change and, further, to educate citizens in their role in order to meet the needs of participative power in complex situations. Thus, a caring Leviathan gives way to direct democracy and the non-centrally organized decision-making of responsible citizenship.

In ecosystems, governance is not the same as central government. If digital ecosystems rely on centralized IT governance they have a tendency to neglect essential controls such as values, language and belief systems. In our present times, governance seems to be a multifaceted term with widely separated ideas and fuzzy concepts behind it. Uyarra et al. discussed governance in terms of 'institutional factors and individual agency in the development and expression of place leadership' (Uyarra et al. 2014). Audretsch linked governance to a concerted management effort of different players in the strategic management of place. He criticized the situation whereby in regional innovation policy '. . .there is no field providing an intellectual framework for decision-making in a manner analogous to the field of strategic management for firms and organizations' (Audretsch 2015).[2]

Managing ecosystems is the management of constant change and transition, and of high levels of complexity. Ideally, managerial intervention must take into consideration complex internal system logics and irrational behavior, because governance of social systems accommodates emerging properties and belief systems. If humans are in the loop, predictable behavior is either based on real action or on assumptions, and relies on smart filtering of big (behavioral) data. This kind of

---

[2]For an overview on governance of innovation see Gebhardt and Stanovnik (2016).

information is needed for intervention and decision-making in complex systems. However, when governance is centered in a *locus observandi* where analytical design, information pooling, decision-making and execution go hand-in-hand it has the bitter flavour of authoritarian control. Mieg and Toepfer therefore introduced the problem of governance as the crucial challenge for social innovation and conceived good governance as a precondition for sustainable urban development (Mieg and Töpfer 2013).

While complexity is necessary for the proliferation of ecosystems, human cognitive capacity runs best on a 7-plus, minus-2 item mode, for decision making. In engineering and artificial intelligence, IT governance and controls employ the logic of cyber physical systems (Mousavi and Berger 2015) and parallel computing to handle complexity. Big data is processed and filtered in real time and the impact of intervention and the outcome are seemingly predictable, even in large and highly interconnected and non-trivial systems when the long term behavior of humans in the loop and their interactions can be projected and the variety of systems becomes a known variable. In this way artificial intelligence holds the aspirational promise of governance of complexity and decentralized power.[3] IT specialists capture interdependences and correlations and the system itself controls the joint learning of engines in the system, and identifies trigger points to influence system behavior like a very talented snooker player—but without the player themself. Flow analytics of movements and developments show when systems become unstable and less resilient. In the 1960s, Minsky's expert systems based on artificial intelligence initiated the support of decision-making in complex environments. Today, cyber–physical systems are used around the globe to 'self-organize' machines and govern the communication and learning process of machines, disrupting global value chains and shifting social systems to different locations and other societal groups. Will the analytical high tech competence help to analyze and to manage the sustainability of social systems? There are three principal questions arising with regard to this scenario. First, can social systems be governed at all? Secondly, what does cybernetic control mean in terms of governance in social systems; and, thirdly, who will take the lead in governance models defining boundaries, rules and IT based intervention schemes?

We stated above that the reality of social systems is stakeholder constructed. The diffusion of concepts and the long-term impact of policies appear to be ungovernable, and brilliant analytical concepts as well as smart filtering of complexity become shallow and mechanistic in the light of real cities and their problems. Above all, fears might prevent a digitalized governance from the very start.

---

[3]In cyber–physical systems even products develop memories and a communication architecture among themselves—see Internet of Things in Wahlster (2013).

5 Humans in the Loop: Scenarios, Danger and Fears in the Smart City

In the light of urbanization and digitalization, change of and within cities is likely to occur. Can this change be managed? Urban and regional innovation studies have discussed the use of high-level technology and cybernetics in smart cities and have warned that cooption of the term 'sustainability' to define the digital ecosystem holds an imminent danger of oversimplification. Film director Terry Gilliam showed, in his 1985 film *Brazil*, that cybernetic-based automated systems can lead to an accumulation of capitalist power where, in this instance, a helpless tenant becomes a victim of his fully automated facility management regime in which different service providers fight for control. Equally, Ridley Scott showed, in his film *Blade Runner* (1982), how global enterprises could rule a high-tech city where biotechnology and robotics merge to unintentionally create a dilemma of social class. Both films—as has much science fiction—paint a picture of dark megacities, nature-free desolation zones and failing disintegrated societies.

In our time the smart and livable city is still a fuzzy concept: sustainability is restricted to reservoirs and upgrading of buildings and districts. Vulnerabilities are visible: the smart service world of the digital ecosystem is centered on the consumer rather than dealing with the environment and spatial entities in addition to the development of civil society. Governance foresees that everything must be organized and the impact times of interventions are being ignored rather than respected. Data driven business models and digital market leadership require new digital infrastructures and platforms, smart apps organize our life, and smart services are transforming leading industries as well as the work–life balance. Will digital ecosystems manifest themselves in cities that reconcile well-being, environmental soundness, high-tech solutions and sustainable business? Is digitally controlled governance good for cities? Who will be in charge of the operations room? The so-called digital ecosystem is a system of new value chains that contributes to sustainability through the reduction of the use of fossil resources, but it will not provide answers to the challenges civic society is facing. Lacking the logics and balance of social and environmental systems, digital sustainability must be seen as a simplification and not a solution. Sennet captures this in his famous statement on the shortcomings of Masdar City: 'Nobody likes a city that is too smart' (Sennet 2012).

References

Adner R, Kapoor R (2010) Value creation in innovation ecosystems: how the structure of technological interdependence affects firm performance in new technology generations. Strategic Manage J 31(3):306–339

Audretsch DB (2015) Everything in its place: entrepreneurship and the strategic management of cities, regions, and states. Oxford University Press, New York

Autio E, Thomas LDW (2014) Innovation ecosystems—implications for innovation management? In: Dodgson M, Cann DM, Philips N (eds) The Oxford handbook of innovation management. Oxford University Press, Oxford, pp 204–229

Berger PL, Luckmann T (1966) The social construction of reality: a treatise in the sociology of knowledge. Anchor Books, Garden City, NY

Deakin M, Leydesdorff L (2014) The Triple Helix model of smart cities: a neo-evolutionary perspective. In: Deakin M (ed) Smart cities: governing, modelling and analysing the transition. Routledge, London, pp 134–149

Florida R, Mellander C, Rentfrow PJ (2013) The happiness of cities. Reg Stud 47(4):613–627

Forrester JW (1969) Urban dynamics. Pegasus Communications, Waltham, MA

Galaske N, Anderl R (2016) Disruption management for resilient processes in cyber–physical production systems. Procedia CIRP 50:442–444. http://ac.els-cdn.com/S2212827116303845/1-s2.0-S2212827116303845-main.pdf?_tid=0b8bea44-a1e2-11e6-b777-00000aacb360&acdnat=1478190471_e2c438f90c23b98a51a52603993f9fb4. Accessed 2 Nov 2016

Gebhardt C (2015) The spatial dimension of the Triple Helix: the city revisited—towards a mode 3 model of innovation systems (Editorial). Triple Helix 2(December):11. http://link.springer.com/article/10.1186/s40604-015-0024-3. Accessed 2 Nov 2016

Gebhardt C, Stanovnik P (2016) European innovation policy concepts and the governance of innovation: Slovenia and the struggle for organizational readiness of the national level. Ind High Educ 30(1):53–66 (Special issue on *Regional Innovation Strategies and Smart Specialization*)

Göllinger T (2012) Systemisches Innovations- und Nachhaltigkeitsmanagement (Volume 10 of Wirtschaftswissenschaftliche Nachhaltigkeitsforschung). Metropolis Verlag, Marburg

Googins B (2013) Leading with innovation: transforming corporate social responsibility. In: Osburg T, Schmidpeter R (eds) Social innovation: solutions for a sustainable future (part of the series CSR, Sustainability, ethics and governance). Springer, Berlin, pp 89–98

Maturana HR, Varela FJ (1987) Baum der Erkenntnis: Die biologischen Wurzeln des menschlichen Erkennens. Scherz, München, Bern and Vienna

Mieg HA, Töpfer K (eds) (2013) Institutional and social innovation for sustainable urban development. Earthscan, London

Mousavi MR, Berger C (eds) (2015) Cyber physical systems. In: Proceedings of the 5th International Workshop, CyPhy, Design, modeling and evaluation, Amsterdam, 8 October. Springer, Berlin

Popitz H (1992) Phänomene der Macht. Mohr Siebeck, Tübingen

Porter M, Kramer MR (2011) The big idea: creating shared value. Harv Bus Rev 89(1):2

Rifkin J (1995) The end of work: the decline of the global labor force and the dawn of the post-market era. Putnam Publishing Group, New York

Ritala P, Agouridas V, Assimakopoulos D, Gies O (2013) Value creation and capture mechanisms in innovation ecosystems: a comparative case study. Int J Technol Manage 63(3/4):244–267

Sassen S (2003) The participation of states and citizens in global governance. Ind J Global Legal Stud 10(1):5. (Article 2)

Schwaninger M (2006) System dynamics and the evolution of the systems movement. Syst Res Behav Sci 23(5):583–594

Sennet R (2012) Nobody likes a city that is too smart. The Guardian, 4 December. http://www.theguardian.com/commentisfree/2012/dec/04/smart-city-rio-songdo-masdar. Accessed 2 Nov 2016

Umpleby SA (2009) Ross Ashby's general theory of adaptive systems. Int J Gen Syst 38(2):231–238

UN (United Nations) (2016) Transforming our world: the 2030 agenda for sustainable development, A/RES/70/1. UN General Assembly, New York. https://sustainabledevelopment.un.org/post2015/transformingourworld/publication. Accessed 2 Nov 2016

Uyarra E, Flanagan K, Sotaurata M, Magro E (2014) Report on the third workshop of the RSA research network on "Regional innovation policy dynamics: actors, agency and learning".

Regions Magazine 293(1):32–34. https://www.researchgate.net/publication/262535423_Report_on_the_Third_Workshop_of_the_RSA_Research_Network_on_Regional_Innovation_Policy_Dynamics_Actors_Agency_and_Learning. Accessed 12 Jan 2016

Vester F (2001) Simulating complex systems as sustainable organizations by transparent sensitivity models. In: Eurosim 2001 Congress, 26–29 June. TU Delft, Niederlande

Vester F (2012) The art of interconnected thinking: ideas and tools for a new way of dealing with complexity. MCB Publishing House, München

von Foerster H (1982) Observing systems (with an introduction by Varela F, ed.). Intersystems Publications, Seaside, CA

Wahlster W (ed) (2013) Foundations of semantic product memories for the Internet of things (Foreword). Springer, Berlin

Weick KE, Quinn RE (1999) Organizational change and development. Annu Rev Psychol 50:361–386

Leading Change in Ongoing Technological Developments: An Essay

Ivo Matser

1 Introduction

Digitalizing is the main issue of this book. In this article, I would like to address the role of leaders and leadership in general how to deal with digitalizing and how to deal with big changes in general. We face many changes because of technological developments and hence, we face many changes in business models of companies, but in society as well.

Digitalizing is a major global development or trend and it creates a different world and therefore many discussions. There might be more distance to employees and to clients, because of "digital relationships", it creates more drop outs (employees who cannot connect anymore in the digitalized world, because of the lack of attention) and in these industries we need less people and unemployment will raise, while the output and the quality will be the same or better/more. But there are more technological trends, so also non digital developments. And additional to these trends, for example the vision on care is changing. It is changing from care to support. So, it is changing that people will be more self-organized and have more self-organized care. From care of people to take care for conditions. We see the same developments in education. Actually, in education we see digitalizing and increased self-organizing in the same time. Students manage their learning processes more and more themselves and educators are becoming facilitators.

In all these cases, companies and institutions need less people to create their output. And, this output might be less "personal" or service-oriented, but at the same time more customized. The good thing about digitalization is that there is less need for scale to remain efficient. In the productions industry we see the same: less people, more robots and more customization and decreasing prices. But in

I. Matser (✉)
ISM University of Management and Economics, Vilnius, Lithuania
e-mail: ivomat@ism.lt

© Springer International Publishing AG 2017
T. Osburg, C. Lohrmann (eds.), *Sustainability in a Digital World*, CSR,
Sustainability, Ethics & Governance, DOI 10.1007/978-3-319-54603-2_8

production industry there is one big difference. The quality and value are increasing while the price is decreasing during the life cycle of products. Manufacturing is running ahead from service or service type (high involvement of human factor) of industries.

In this article I do not want to solve this problem. I want to address leadership and leadership dilemma's while we create our own problems and stick to the past and hesitate to connect to the future.

2   What's the Symptomatic Problem?

The biggest disadvantage of digitalizing is the decrease of employment of people, it is stated. Seen as a subsystem, I would agree with this. This means increasing unemployment, less buying power, and problems in society because of the increasing unemployment. People will be less connected in society. Another important development is the increasing minimum level of competences and pressure of many jobs. In many industries the blue collar workers became white collar workers. The result is an increasing percentage of drops outs; people who are not able to develop the higher level of competences and/or are not able to deal with higher pressure or even to high expectations of the degree of independency at a workplace.

In the subsystem this is no rocket science, it is more or less predictable. But, the economy is a complex system. A complex system may be defined as follows. In the first place the system consists of many subsystems. These subsystems are independently active and they influence one another. Secondly, the subsystems are "learning" systems, which means, they use information of many systems and they provide information to other systems. Thirdly, the subsystems are adaptive, so they move and influence at the same time.

So, maybe the biggest problem of thinking about the influence of digitalizing as mentioned before is that we do not understand the system. Maybe we are able to understand a subsystem as explained above, but this does not mean that we have the big picture. And maybe we face resistance and resist ourselves important developments, because we do not understand and we try to keep many things as it was before.

3   What's the Real Problem?

The main problem we face is that many people resist many developments. We try to stop developments we do not understand, because we try to understand them from the past. So we do not use developments. And there are many developments we can never stop. There are many technological developments, as digitalizing, DNA-technology, 3D printing and Nano-technology. They will continue and they will bring many good things and many improvements in products and in sustainable products and sustainable production methods. For products as well as for services. In the end we need less people, less logistics, less energy to create better and cheaper output. So, what's the problem? I would say, let us be more optimistic. Unfortunately many people are pessimistic because of the fear of unemployment etc. Maybe because we look backward too much because it is difficult to understand the future developments.

Before discussing leadership issues, let me share, why we should be optimistic. Because of many technologies as mentioned before there will be many improvements in health care, distributions of medicines, sustainable life cycles of products etcetera and it will become cheaper and it will have a better footprint. So people will have buying power for other products, same products with more features and services. Some industries will gain many opportunities. For example, the leisure and tourism industry and even the care sector will have many opportunities and possibilities to grow, to innovate and to create a bigger industry and more employment. These industries are very labor intensive and will create many jobs. So, other subsystems will react and will adapt to all kind of developments. We even do not need to manage it; it will happen. But, it is very interesting that many people worry about the digitalization and the labor market and do not really think in opportunities, like leisure, sports, health care and welfare, as mentioned before.

What is really bad, if we try to resist and not to adapt and not be entrepreneurial or optimistic? And therefore try to manage it on the wrong direction and not to use it. Because if we resist and is we are frightened, we will have only a short term perspective and this will lead to bad decisions. In these cases we try to keep industries in a traditional way, will be to expenses and supported by governments, all kind of funds and this will detract money from the economy. Very expensive and useless, because in the long run technological developments and their impact will always win.

Let us define the main problem: why many leaders are not capable to lead change in inevitably and major trends and mostly are resistant to change themselves. I will not address all kind of psychological issues, I will focus on business logic and the logic how people behave, based on my own experiences. During my career I have been leading many companies/institutions in high human involved service industries and the reason I was leading it, was because of the need of change.

4 Leadership Dilemmas for Change

Leading change is the most important purpose of managers. Management as repeating routine is not that difficult, but this way of management will disappear. Market dynamics and increasing complexity will cause many changes and organizations (private, public, companies, and institutions) have to adapt. So management is all about leading change. Many managers have problems with leading change because they face all kind of dilemmas.

The first dilemma is the dilemma of control and to explore new future business. In many organizations we developed all kind of systems to be in control. All kind of balanced score cards, cascading objectives and goals, forecasting the future and plans with targets and predictions. So, we try to manage the future and similarly we try to find solutions we do not know yet. Sometimes managers try even to plan innovations with targets. Very naïve, and we know the success of less planned creative processes. But, in general, many managers want to be in control. Mostly, because they are obliged to report to their managers the same way. It is very interesting to observe that people are aware of the uncertainty of the future, but also plan as if they are able to predict the future. Maybe it is even easier to predict the far future, because of megatrends and more difficult the near future, let us say from now to 5 years.

The second dilemma is the dilemma is change and stability. Mostly we think and we communicate that after a period of change, there will be a period of stability. This stability is perceived as no change. If people complain in their organizations they mostly complain about clarity, communication and too many changes. We create expectations about stability and change. In fact, the periods of change last longer and the periods of no change are becoming shorter. It seems more and more, that periods of no change are becoming moments instead of periods.

Thirdly, there is a dilemma of old culture and create a new culture. This is probably the most difficult one. All those plans about culture change. Escaping one the old one and trying to find the new one. Long lasting processes and mostly low impacts. What is the problem here? The most sustainable asset of each organization is culture as the collective actual and historic traditions, rules and behavior. It is more "fixed" than any structure or system. It is not tangible, but really fixed. So, in perceptions of employees, this is the most certain and most save, even if they do not like it. So communicating about culture change will create huge resistant instantly, visible and below the carpet. The inside world became extremely unsafe for people and they got stuck.

The fourth dilemma is that leaders think that people do not want to change. We have the pre-conditions that people do not want change and do not want to change. This is the reason why we position change mostly as problematic. To avoid the situation that all people see change as a problem people communicate it in many ways. Sometimes, we create a problem with urgency and importance to force change. This will be perceived as negative and it will not be perceived as invitation to future activities. Sometimes, there is no problem, but we have a long term need and we draw a sort of "heaven" for people and they think one became mad or at least very naive.

The fifth dilemma is top-down and bottom up. Do we have to manage organization in a more bureaucratic way? Top down and obeying systems and procedures. Many people like it, because it feels comfortable and it is perceived is "being in control". Or should we be more bottom up? Sometimes confused with the idea of democratic processes. As if we all are experts and educated as decision makers. But bottom up makes sense. It is a way to involve people better and if people feel a kind if ownership they will be more motivated.

The last dilemma is about trust. Do we have confidence in employees, clients, suppliers or not? Many systems, for example employee assessment and rewarding systems are based on mistrust. That's also the way how we deal with external stake holders. Interesting is that mostly we like to trust other people and we also liked to be trusted. But somehow, we often do not allow to trust. Maybe it is based in the idea that humans are one of the production factors in economic models, on the same level as materials and land. It comes from the period that we see sharing as having less for more people and currently sharing is more related to multiplying.

5   What to Do with Dilemmas?

Many managers have this mindset of dilemmas. Mostly we think in opposites. Like yes or no, like black or white, like for or against. To face dilemma's we try to find a way or we negotiate a compromise. Sometimes in the middle and sometimes more to one of the opposites. Of course, to find a compromise might be a huge achievement, but mostly no one is happy with a compromise. In many cases it feels like "stuck in the middle".

Let us try to find another way to escape from dilemmas and let us forget this way of linear thinking. It is really too simple to think in opposites or in only two best options. Everything will be stuck in the middle and no progress will be made. Maybe we have to try to use both opposites and even make the difference between the opposites bigger. So, very black and very white and let us try to find new colors. Or maybe in business language: let is find synergy between the opposites and let us use differences. Similar to diversity. We know diverse teams create better and more unexpected results.

The best thing to do with a dilemma is to stop thinking in dilemmas. Below I will explain some ways to go based on the six dilemmas, mentioned before. The next paragraph is about change as a process and then about how to organize.

6   Leading Change and Do Not Care About Dilemma's

Learning
It may be too obvious, but understanding what is learning it is the best to look at kids. Learning how to walk is not somewhere in between sitting and walking (as compromise of two dilemmas). It is trying, experimenting and if kids succeed, they even try to run. So, make the next step and then look further. Managers often think in gap-analysis, where we try to bridge shortcomings to a construction of a wished reality. It is a mindset of having problems, because of these weaknesses. I would prefer the approach of stepping stones. Take the next step, learn and prepare

the following step. From the rational or logical perspective one could argue the true differences. But, there is a huge difference in mindset or mentality. And it is not that difficult. It is not easy not to learn. We see similar things in the way how entrepreneurs act. The causation method is based in gaps and milestones, the effectuation method is based on learning and stepping stones. In trends like digitalizing it will become inevitably. Because, many business activities will be more interactive and the wall between inside the organization and outside the organization will dissolve more and more.

Change as Business as Usual

Secondly, we mostly strive for stability. And having a change we promise there will be stability after the change. We know better. We know life is about change and business is about change. But, we feel change as unsafe, because we promise stability. So, we create expectations, maybe because we care about people and want to reassure them, but, in fact we create very unsafe conditions. Sustainability is based on change; let us remind Darwin about the human species. It is similar to what many people think about entrepreneurs. We see these people as more risk taking people. It is the opposite. They are more connected to the future and no change or mostly looking back is very risky. So, is we see change as business as usual, we will find it more easy to communicate about it. Then it will be more common that our communication about change and development will be very clear and as transparent as possible. Because that's what really frightens people: a lack of transparency and clear message (also about what you do not know).

Culture and Change

So, if change will be as business as usual, it will become complicated to talk about culture change, because a continuous change of culture will kill any organization. Culture, defined as the collective attitude, behavior and traditions is what people connects to each other and to the organization. Even, if people complain about the culture. I strongly doubt the relevance of any culture change project. Culture is by far the most sustainable "thing" of any organization and is the basis for happiness, success and whatever kind of mentality. Only using the word culture change will create resistance instantly and you lost the game of influencing people towards future activities. Of course, digitalizing and trends from care to support will ask different attitudes and behavior of people. Call it learning by doing. And maybe later, one will say, maybe our culture has been changed slightly the last years. That's how it works.

7 Conditions for Continuous Learning and Change and Not Being in the Trap of Dilemmas

The Maturity Level of Being Self-Organized

Try to use the hierarchy as less as possible. Of course we need formal decisions and an organized decision making process. I do not believe in maximizing self-organization. But, there should be a sort of maturity level of being self-organized, which means that the organization does not need the hierarchy of day to day operations of the routine planning processes. This is conditionally for teams to be valuable and to organize customer focused processes. Because in many cases the formal organization does not represent the processes. There are many management methods regarding maturity levels. The mid-range of maturity levels mostly are system oriented, which means that the internal processes are connected, consistent and well organized. Above the mid-range of maturity levels, the systems are anchored in the outside world. So, the real business world should not be in the dilemma of bottom up or top down and finding a compromise. Or negotiating the influence and power in matrix structure. No, being in a mature way self-organized makes the formal structure lean and more effective and a healthy formal organizations supports the level of being self-organized.

Management by Example

Be the change yourself, behave like you want that people want to behave. In a supporting and learning organization and in self-organized companies there is no place for ego type of leaders or narcissistic leaders. No leadership by fear or power anymore. The meaning of strong leadership will change. From paternalistic leaders to visionary and people oriented leadership, motivating people to learn and to experiment. Maybe, in the future the important leaders will not become gurus anymore. In digitalized company, mostly international, the question is about leadership in technology and how to create virtual international teams. In this case, the conditions for people are the most important and significant for competitive positions of companies. Then true leadership is mostly supporting people and leading the journey to the future, (step by step). Having trust or no trust, was also mentioned as one of the dilemmas. It may be obvious that the more supporting way of leadership is based on trust and creates trust. Even more: it will influence positively the value of the company and the value of the complete value systems of suppliers and clients, because cooperation in the system is needed for value creation. So, management by example based on trust is as well as important for the internal organization as to the external stakeholders as suppliers and clients/customers.

8 Back to the Future Technological Developments

Technological trend is the main reason to think about issues as leadership and change. It does not even matter, what kind of technological development. Because many technological developments will change the world and therefore business models dramatically. Digitalizing is only one of the trends. But many trends will result in customization, extremely high quality of products and services, and low footprint. Maybe the biggest problem is big data. We need many, and we have too many. Are we able to invent algorithms and computers to calculate (quickly)? So, the directions of new business models is more or less the same in many technological changes. There is really no reason to resist, because it contains in general many improvements for society. It is better to support and to adapt.

9 Conclusion

Don't worry, be happy. Embrace digitalization and new technologies in general. The most important is getting rid of many dilemmas created in the mechanization period of our economic history. We did not even solve these problems during the information period. It seems that many manufacturing companies understand this already for many years. For service oriented organizations, with high involvement of people there are many opportunities. And people who like to work directly with other people and clients, will find their way in new business of leisure, care and social entrepreneurship. A helping hand from government and/or a role for the public sector might be considered to help to change the big picture. Forget the dilemma's, do not try to solve them anymore is my main message. And it should be supported by renewed and true leadership. And there will be space for many new business with intensive human capabilities in industries and sectors I mentioned before. And having more social entrepreneur in healthcare and welfare is a very relevant for social sustainability.

Part II
Markets, Business and Stakeholders

How Digital Reframes the Business Case for Sustainability in Consumer Markets

Alexander Holst, Christoph R. Löffler, and Sebastian Philipps

1 Digital and the New Business Case for Sustainability

Sustainability has successfully shaped debates, politics and even the lifestyle of a small portion of individuals. However, for mainstream business in consumer goods, utility, services and retail, sustainability remains a peripheral and not core to their go to market strategy. More than 90% of consumers in Germany find the deployment of renewables important or very important (Renewable Energy Agency and TNS Emnid 2015). At the same time, only around 20% of consumers actually buy energy from renewable sources for their homes (Statista 2015). Where does this 70% gap stem from?

Following up on this question, we first cast light on the consumer perspective on sustainability today, and describe the 'split personality' we observe with many consumers. In a second step, we lay down three prerequisites for overcoming the split: sustainable products and services should personalize sustainability, they should seamlessly fit consumers' lifestyles, and should be conveniently accessible. In a third step, we explain how digitization makes the provision of these sustainable products and services much more affordable and manageable for companies today.

A. Holst (✉) • S. Philipps
Accenture Strategy & Sustainability, Accenture GmbH, Friedrichstraße 78, 10117 Berlin, Germany
e-mail: alexander.holst@accenture.com; sebastian.philipps@accenture.com

C.R. Löffler
Fjord – Design and Innovation from Accenture Interactive, Pappelallee 78, 10437 Berlin, Germany
e-mail: christoph.r.loeffler@fjordnet.com

T. Osburg, C. Lohrmann (eds.), *Sustainability in a Digital World*, CSR, Sustainability, Ethics & Governance, DOI 10.1007/978-3-319-54603-2_9

The 'Split Personality' of Today's Consumers

What is the root cause of the empiric gap between wanting energy to be renewable and purchasing renewable energy? Is it a specific German phenomenon? It is not,[1] but the gap between expectations and reported actions is particularly stark for consumers in developed countries as compared to those in developing countries, according to a study by Accenture and Havas (2014).[2] Here, we will explain how this assessment underlines our suggestion that many consumers display a split personality when it comes to sustainability, and what it means for companies when they ignore this split of personalities.

Sustainability is a feasible term for framing a general claim for companies to do better for society. Overall, dissatisfaction with the sustainability performance of business persists on a global level, with more than 70% of consumers demanding businesses to act more responsibly according to the above study. Unsurprisingly, consumers want a better and more sustainable world.

Looking at consumption, however, it may not be sustainability itself that sells but what consumers take it for in their very personal context. For a Chinese consumer sustainability may for instance stand for food safety (Brookings 2016) or a Western lifestyle. For some German consumers it may mean fair pay for cacao planters in Latin America. Many more Germans, however, may perceive it as an abstract concept adhering to which involves mastering of a myriad of labels.

We propose that consumers carry around both the general claim for a better world coined a 'sustainable' world, and very specific consumption needs that may or may be not sustainable. The *idealist protester* and the *pragmatic buyer* coexist in most of us. Their coexistence leads us to diagnosing a split personality, albeit not in the medical sense. We do not speculate here, to which extend people suffer under this cognitive dissonance. We do however presume that most of them prefer products and services representing their *individual* expectations toward a better i.e. sustainable world—if they find such products and services on offer.

For companies this preference is a material business potential. In emerging markets this notion already manifests itself in business strategies. Metro China has understood and implements a safety centered framing of sustainability (Starfarm 2017). Also in developed country markets companies seek to address consumers more specifically. One example, to stay in the food market, is the REWE Pro Planet label explicitly stating, which product related hot spot has been mitigated, for instance promising fish was caught in a way that sustains fishing grounds.

How big is the potential exactly? According to Accenture and Havas (2014) only 21% of responding German consumers reported to already consider sustainability when making purchase decisions. At the same time 88% expected companies to be more responsible. This equals a gap of 67% of consumers not translating their

---

[1] Also, the phenomenon is not limited to energy markets, although their nature may make them more prone to create it: In fact, what comes out the plug socket cannot be discerned as sustainable or unsustainable in itself. It remains electricity.

[2] The study is based on a sample of 30,000 consumers.

expectation into purchasing preferences. In contrast, for the sample of Chinese consumers the gap was only 26%, given a relation of 44% reporting to consider sustainability and 70% expecting more of it from companies.

Why does business not manage to reach out to the 67% of consumers in Germany who should be willing to opt for more sustainable products and services in their individual sense when given a choice? Instead of turning to theory, we illustrate the case for a fictitious German consumer we name Klaus, applying a persona based approach.

Klaus is a typical German middle class man in his early forties. He lives in a medium sized town not far off an industrial region. He has a wife, one kid, and runs his own small business. He buys organic milk but cheap deep frozen poultry. Since two years, Klaus has been using the word sustainability more often: Back then, he became part of a local protest movement against the construction of a coal fired power plant nearby his town.

His focus on the negative impact of coal fired energy made him pay greater attention to local air pollution and climate change in general. This directly affected the purchase decision for his new car, and the upgrading of the air filter systems in the little workshop that belongs to his company. And, it made him change his electricity provider for both home and business. For Klaus, sustainability means low carbon and low emissions, and this is what Klaus stands for in his circle of friends.

They all, including Klaus, do not know that in fact, Klaus has some sort of a split consumer personality: Despite his claim for a coal free economy his private pension scheme is highly carbon intense, investing large amounts in South African coal fired power plants. The input materials he buys in larger quantities for his workshop come from East Asian suppliers with lowest emission standards, and even his car is not as clean as he expects. Klaus simply lacks the time and knowledge to manage the complexity of these relevant levers he controls—not to mention the cumulated impact the myriad of daily shopping decisions he takes on the go.

Given that even Klaus the anti-coal activist fails to consider sustainability in his very narrow sense, how difficult will it be for an average consumer to consider sustainability put in more comprehensive terms? What would Klaus do, what would consumers demanding more sustainability do if all this became much easier? What could companies earn by reducing complexity? What does it take to end this waste?

What It Takes to End the Waste

How can we cope with Klaus' split personality? We assume that consumers are ready to consider and willing to pay for sustainability when considering three aspects. First, products and services need to address the individual values of consumers, i.e. sustainability needs to be personalized. Second, products and services need to seamlessly fit the individual lifestyles of consumers. And, third they need to be conveniently accessible. These prerequisites sound straightforward but they prove challenging to implement.

At Accenture Strategy, we have helped many businesses to 'personalize' sustainability. We identify the value that sustainability can create for a specific company, and we even quantify this value in terms of revenue increase, cost and risk reduction, and the improvement of intangibles. Our clients can operationalize this knowledge strategically and within their operating models to increase overall business success.

Personalizing sustainability for consumers is, however, a different case: While companies look at revenues, costs, risks, and intangibles, one and the same consumer looks for a variety of things at different points in time, ranging from purpose in life to satisfying ones thirst. While companies consider their profit and loss statement, consumers follow their feelings. Consumers simply bring an implicit need to the point of sales. This demands new ways of linking the point of sales with the value chain.

Recalling Klaus bad investment decision—how could his investment advisor have personalized sustainability for Klaus? Offering Klaus a sustainable investment option would have been one step in this direction, with some probability of Klaus buying it simply because it considers his grim toward coal-fired electricity. However, what about offering Klaus a low carbon coal free investment product?

Being able to personalize sustainability for a specific consumer group is only the first step. The second prerequisite is to understand when to offer which product or service to a consumer, given that having on the shelf everything at any time in conventional ways is inefficient. This is difficult enough in marketing products such as for instance chocolate bars. It becomes even more challenging when companies seek to address subtle ethical values of a specific consumer at a specific point in time. Anchoring products and services in individual lifestyles is a way of achieving this. We will elaborate how this may work using the concept of living services by Fjord and Accenture Digital at a later point.

Using the example of Klaus: His investment advisor would not only need to have a low carbon coal free product on offer but she would also need to know about Klaus personal coal vendetta to offer him exactly what he needs at this point in his life.

The third prerequisite is accessibility and convenience. The Global Consumer Pulse Research by Accenture (2014) found that even for the formerly predominantly price sensitive German market convenience of transaction has become the top driver of customer satisfaction. Looking at sustainability, this explains the unwillingness amongst the majority of consumers to study dozens of labels or read through lengthy background documents when making purchase decisions. Even more straightforward ways of presenting sustainability such as the mentioned REWE label by design fail to conveniently offer information on an individual basis. What if Klaus personally does not care about fishing grounds but about labor conditions for fishermen?

When it comes to Klaus investment decision, what is more convenient: looking for an investment advisor at a bank who offers coal free investments, or proactively being offered a selection of low carbon investment opportunities by a third party?

The example of Klaus illustrates how conventional businesses may find it hard to offer personalized products to individual consumers at the right point in time in a convenient way. No surprise that the 67% have remained untouched so far. But?

How Digital Is Creating Mainstream Sustainable Products and Services

Tapping the business potential of sustainability appears in new light when considering how digital disruption is currently reframing consumption and production. This is, digital technology not only becomes more sophisticated but also more affordable—and it dominates the everyday life of an increasing portion of consumers. All this affects the opportunities for companies with regard to tapping the full potential of sustainable consumption and production, and it reframes the competitive relevance of sustainability in the mainstream market for consumer goods.

Digital has become much more sophisticated since the time when desktop computers were first connected to the internet via cable. Technologies such as mobile, Internet of Things, social, the Cloud, and analytics evolve quickly. Their combination opens up immense opportunities to personalize products and link them up with services. One example from daily life is the toothbrush. Thanks to an App and a Bluetooth connection the brush developed by Oral-B could for instance monitor Klaus brushing behavior and style, and advise him on how to improve (Oral-B 2016). Just imagine how difficult it would have been to provide a similarly personalized service fifteen years ago—not to mention how expensive.

Digital technologies have not only become more sophisticated but also much cheaper. Cloud services are but one example with a prices dropping by a quarter between 2012 and 2015 according to the Economist (2015). Thanks to such price trends, today, lean startups can challenge corporations. For instance, financial technology ventures so called fintechs are putting into question established business models in the finance industry. Investors apparently see the potential, given that investments in fintechs tripled in 2014 to reach a total volume of over 12 billion USD (Accenture 2015). One of the success factors behind fintechs is their orientation toward making the consumers' lives simpler, according to Accenture Strategy (2016). Thinking of Klaus, this is exactly what he would have needed regarding his own coal free investment decision: an easy to access and convenient choice.

Thanks to digital technologies being more sophisticated and cheaper, opportunities to create and combine new products and services are abundant. Equally important, the increasing dominance of Digital in most consumers' daily lives brings about endless numbers of potential touchpoints between the consumer and new products and services. In 2014, mobile became the primary access point to the internet in the United States according to ComScore (2014). Klaus is not a digital native but nevertheless his mobile phone dominates his private life. To mention few standard applications, he takes and shares photographs, engages in instant messaging, checks on his tooth brushing behavior, shops, listens to music. Obviously it does not stop here. With new interfaces between man and machine being developed, Klaus' daughter Gabi will grow up in a world where some sort of artificial intelligence will always be at her side waiting for words, gestures, and other forms of communication to pick up on. Even the smallest interaction will add to her digital consumer profile, which tracks her personal values and preferences. Based on this profile it will be easy to automatically personalize sustainability for Gabi.

2   A New Value Proposition Behind Sustainability

Businesses need to understand how to navigate this emerging landscape with a strategic sustainability angle. Those that manage will be rewarded by stakeholders, consumers, and shareholders alike.

When You Are Ahead of Your Stakeholders

In future, not only consumers will be less anonymous—companies will be, too. The Cloud, mobile, social, and other technologies put stakeholders along the value chain in the position to detect sustainability issues, share them amongst each other, and effectively exert pressure in consumer markets. Companies need to stay ahead of this development by adopting advanced traceability, transparency, and trust strategies. Otherwise they face regulatory risks, loss of revenues and brand value, and additional costs.

Stakeholders have long begun to use technology effectively. The Kit Kat campaign by Greenpeace is but one example for an approach that links upstream sustainability issues with end consumer behavior using social technology. In the video that went viral, an office worker chews on an orangutan finger baked into a Kit Kat bar. Blood dripping down his chin translates negative effects of palm oil plantations on the orangutans' habitat into emotional language understood by netizens. Even to Klaus who focuses his concerns on the coal fired power plant next door, eating an orangutan finger would not appeal, and make him think about palm oil—at least for a while.

The value at risk for companies is considerable. On behalf of a chemical industry company Accenture assessed the profit at risk connected to traceability along the value chain. The research linked traceability to regulatory and statutory levers, market and consumer response levers, supply chain benefit levers, and recall and risk management levers. For the analyzed company, the profit at risk connected with these levers added up to 300–350 million EURO, in 2012.

In a digital world, technology backed sustainability strategy will enable companies to stay ahead of stakeholders when it comes to making additional regulation unnecessary, reducing profit reduction due to scandals, and avoiding other costs such as expensive recall and take back schemes.

When You Can Offer Your Customers What Really Matters to Them

Defining the right value proposition remains a key prerequisite for business success. Personalizing sustainability is a way to enhance the value proposition to individual consumers based on their world views. In this sense, sustainability will matter more than before; and offering personalized sustainability that seamlessly fits consumers' lifestyles in a convenient way will be part of the core business for many.

Sustainability matters to consumers in different ways. Depending on who they are, some consumers may buy what is traceable, others may pay for what subconsciously makes them feel in harmony with their inner values, a third group may use sustainable products and services for what they offer in terms convenience or other non-sustainability performance aspects. Catering to all of these three groups becomes easier for companies that understand individuals and their lifestyles in real time, and invest in addressing them holistically. Think about Klaus and his investment decision. Proactively offering him a zero coal saving scheme would not only help Klaus to overcome his split personality but would, in turn, also remarkably enhance the value proposition behind the financial product—with an effect on Klaus' demand or his willingness to pay or both.

Google's advertisement business model is based on translating individual search and surf behavior into more targeted product offers. However, how many companies that use google ads use agile ads that address the viewer based on his personal values? Would one and the same financial product be offered to Klaus as zero coal choice, while the person next door is offered it as low risk choice at the same time? Would their search and purchasing behavior have direct effects on fond management and product development?

In a digital world, opportunities go far beyond reaching out to the right people with a product or service. In a digital world, technology will enable companies to adapt the presentation, development, and pricing of products to individual conditions in real time. The algorithm may not only know which product to feature in the browser but also *how* to present it—and how to price it, based on the customization. One example would be to offer different versions of one and the same shirt to different users, not only in the right color but also in an organic, fair trade, or circular materials version. With the right technology behind it, sourcing of input materials could be steered by views and purchasing behavior in real time.

The shirt example illustrates how technology can also work as nervous system linking consumers with the most upstream parts of the value chain, eventually making supply chains more circular. Eco-ATM or similar services for recovery of communication equipment run by Baidu are good examples for the circularity dimension. They organize cash offers for used communication equipment and thereby make it much more attractive to, for instance, feed an old mobile phone back into the system. They carry the potential to send direct price signals from commodity markets—using rare metals from disposed phones—to the downstream end of the chain, i.e. Klaus wondering what to do with his old smart phone. What if they also knew how to address whom using what argument? Klaus may be even more interested in the carbon effects of giving back his phone.

When Investors Understand What Your Sustainability Is Worth

Shareholders will appreciate sustainability driven improved risk management and marketing strategies in their investees core business. Already today, investors and corporations acknowledge the relevance of sustainability as route to competitive advantage. Digital brings opportunities to substantiate this link.

A study by Accenture and the UN Principles of Responsible Investment found that 88% of investors, and 79% of CEOs regard sustainability as a route to competitive advantage. The same study finds that 57% of CEOs think they already capitalize on sustainability related opportunities strategically, while only 9% of investors agree on that (Accenture and UN PRI 2014). Investors traditionally relate sustainability to risk, for instance in the form of temporary revenue decrease and legal costs in reaction to environmental or societal havoc wreaked by an investee. However, with sustainability being personalized it becomes more important for the value proposition and long-term revenue perspective of invested companies.

The German utility industry illustrates how sustainability trends can impact business models and share prices in an industry once deemed as boring but reliable investment target. As one example, E.ON in 2016 separated their conventional generation capacity under the umbrella of UNIPER while focusing their core business on renewables, grid services, and other energy related services. Eight years of decreasing share prices had preceded this decision (Frankfurt Stock Exchange 2016). Initiatives such as the Asset Owner Disclosure Project (AODP) drag investors themselves into the spotlight with regard to how sustainable their portfolios.

Digital not only enhances the relevance of sustainability for investors, it also makes it easier to operationalize sustainability aspects in investment. As suggested above, we believe Digital will empower stakeholders and consumers to detect and communicate shortcomings in the environmental and social realm along the supply chain. Beyond that, digital responsibility itself will become a sustainability issue. According to an Accenture Strategy study (Accenture Strategy 2016) digital technologies disrupt how business leaders and organizations establish trust and ethical behaviors, while 83% of the respondents to the underlying survey say trust was a cornerstone of the digital economy. On the operations side, digital technologies will enable more transparency and can facilitate common metrics used across supply chains and by investors and investees alike. CEOs and investors named such common metrics as relevant step toward closing the gap in sustainability investments (Accenture and UN PRI 2014).

Returning to Klaus, he might buy a coal free low carbon investment product the next time. He would likely buy it from a provider who understood how to address his personal views and values. He may even decide to sell his existing carbon intense portfolio. If he is not the only one doing so, the provider of the conventional investment products may see a need to rethink their investment strategy or at least become better at monitoring carbon and other sustainability aspects throughout his portfolio.

3   Emerging Sustainability Business Models Driven by Digital

Digital reframes the business case for sustainability. Digital technologies enable the personalization of sustainability, a seamless fit of sustainable products and services in the daily routines of consumers, and convenient access to these products and services. The emerging dominance of Digital in today's lifestyles facilitates scaling up of the resulting sustainability related business potential. People like Klaus do not change in essence, but their willingness to pay for specific social and environmental aspects comes into reach of companies. A variety of business models evolve to tap this potential. We reckon that rethinking strategy with a strong technology angle and a profound understanding of lifestyles is key to making these business models thrive.

Business Models for Sustainability in a Digital World

Business models and the potential they carry change with digitization. We argue, this particularly holds for sustainability related business cases. We have described that individual values and beliefs around sustainability are often subtle and complex. These circumstances make sustainability related business models benefit from the availability of big data and respective data science technologies. It becomes affordable and doable for companies to uncover individual sustainability preferences and address the respective willingness to pay. Regardless of whether they approach it alone, with other businesses, or in cooperation with consumers, data and digital technologies are key enablers to the sustainability business case in the digital era.

Companies can, on their own, monetize data concerning consumers' sustainability related values. With Klaus consent, any company that owns a broad data set on Klaus can sell his data on to those who want to target Klaus individual sustainability related values and willingness to pay. For instance, an investment broker could buy data and use it toCorrection Update detect, that she can sell Klaus a zero coal financial product. Instead of selling raw data, the (Hofman and Spijker 2013) data owner could also offer intermediary services to others, providing them with dynamic profiling for Klaus.

Alternatively, companies can collaborate more closely with their suppliers to offer innovative customized services. For instance, they can ensure supply chain transparency, and operational sourcing practices needed to tailor products to individual consumers at affordable prices. In addition to this vertical approach, companies can enter into cross industry partnerships, creating value networks that jointly offer matching services and products to one and the same consumer or consumer group. In our example, the investment broker would have benefitted from knowing Klaus' interest in renewable energies. It also would have been easier for the broker to offer a coal free financial product when having direct access to portfolio data of providers of financial products. Such collaborative models frequently depend on sharing of data, and cross fertilization of product design and operating systems.

Beyond these B2B models, companies can directly work more closely with consumers, for instance, setting up sharing economy business models around the idea of using consumer data to function as matchmaker. Often, shifts toward product as a service models also involve close interaction or cooperation with consumers. For instance, Deutsche Bahn initially relied on users of their bike as a service scheme to tell the system where the bike was parked.

These are only a few examples on new business models with a strong sustainability angle. They all have in common a strong reliance on digital interfaces, individual and aggregate data on use patterns, and large scale data flows.

Rethinking Strategy with a Strong Technology Angle

In the digital era, strategy and technology converge. The business case for sustainability and related business models exemplifies this. Strategic considerations with a strong technological angle lie behind many of the above described business models for sustainability. We want to highlight three considerations here.

First, companies need to get hold of their supply chain. Increase transparency, traceability, and steering capacity. Embracing this can earn them new opportunities, and secure trust on the side of customers, stakeholders, and investors, as described above. First companies such as 'honest by' (Honest by 2016) aim for full supply chain transparency on product level. Still, transparency and traceability are a challenge for many companies today, as organizational barriers remain. However, the technology bar has been lowered remarkably. Mobile enables the real time tracking and steering of goods from any location. The Cloud can host platforms to share respective data. The Internet of Things supports automation of supply chain communication across companies and with equipment used by consumers.

Second, companies need to think circular. The consumer is not the downstream end of their value chain but a crucial node in the circle that will enable them to reduce waste and create new value. Business models such as product life extension, sharing, product as a service, circular supplies, and resource recovery can help avoiding 4.5 trillion USD until 2030 (Lacy et al. 2015). The consumer is a crucial part of these business models in many case, thanks to mobile and social technologies. Big data analytics, Internet of Things, and the Cloud round up the array of digital technologies that promise to make the circular economy a paradigm shift.

Third, companies need to embrace consumer expectations. Instead of categorizing consumers upfront and addressing them with marketing, companies can cater to consumers like Klaus in their current situation, provided they can harness the big data available and translate it into personalized offerings at the point of sale. Social and mobile, together with big data analytics offer opportunities for customization at scales unheard of prior to the digital era. However, rethinking the consumer takes more than the application of technology, as the following and last section indicates.

How to Offer What Really Matters to People

A common misconception among consumers is that choosing a sustainable product translates into a certain compromise in quality or in the experience of using the product. Conceding quality, price value, and user experience for the benefit of environmental friendliness has indeed been a reality for many consumers. One just has to remember the complexity in charging first generation electric vehicles after they didn't get you very far, or think about many of the current generation smart home solutions that are so difficult to install and operate that they appeal only to technically-savvy and environmentally conscious pioneers, by definition a small minority of the population. Too many sustainable products are designed with a high level of tolerance for compromise in quality and user experience, and are not being bold enough to aspire appealing to mass-consumers rather than an environmental conscious minority. Compromising for the sake of environmental friendliness is not a preordained causality for brands, products and services, and today, there are more and more examples of well-designed products emerging that prove this point.

Owning an electric vehicle today in a city like Berlin, paints a stark contrast to that consumer experience even 2 years back. Tesla, with its supercharger stations is making important contributions to close the experience gap to a well-functioning, well-established carbon-fuel gas station network. They realized early on, that providing a user-experience with no compromises to its owners, would require the existence of a faster and more accessible fueling experience. With the Model S and X, however, Tesla's biggest coup has been to design products that do not represent any compromise in vehicle usability, and may arguably even be an upgrade, when compared to the product experience of their carbon-buzzling competitors. With this, Tesla has revolutionized the image of an electric vehicle, lifted themselves completely outside of a niche environmentally conscious target population, and managed to appeal to a very broad, albeit luxury, consumer segment. With the Model 3, they are breaking the final stigma of electric vehicles, that is the price value equation, and making this model affordable to a mass-consumer market.

The North American retailer Target has also managed to take sustainable mainstream, with the launch of their "Made-to-Matter" product line. Thousands of products and brands are scored on a simple point system covering factors such as ingredients and packaging, and labeling these in their stores with easy to digest, well-designed information for shoppers. The next generation of such product labelling may well describe to consumers the precise information that will convince them to choose a more sustainable product over the cheaper alternative.

What a highly successful electric vehicle brand and a mainstream retailer will then have in common is that they firstly, both sell products and services that were well designed, centred around mainstream consumer needs, and secondly, that they disrupt through digital services. Human-centred design, analytics and digital technology is enabling these companies to empathize with needs of their customers, understand their customer journeys and use-cases that start and end well beyond the consumption of the actual product, and based on this design and deliver focused product solutions, or digital services wrapped around their products that match precise consumer opportunities.

Fjord and Accenture describe well designed digital services that seamlessly integrate into customer routines and experience journeys as Living Services. Living services wrap around us, learn about our needs, intents and preferences, so that they can flex and adapt to make themselves more relevant, engaging and useful (Fjord and Accenture 2015). By definition, Living Services do not represent a compromise in customer product experience, but on the contrary, they will likely enable a richer, more beneficial and elegant experience of a product or service. Recall the smart home example, and the fact that smart home solutions adoption is extremely low today, while the potential benefits to consumers of a well-designed (aka Living Services version) connected home solution, as recognized by Google and Apple through their entry into this market, will be immense. In these next generation smart home solution, sustainability will be a powerful side effect, but it will not be central theme around which these services are designed.

In the coming years, we will see Living Services start to transform a wide range of industries and markets. And as Living Services wrap around products and retail experiences, addressing sustainability related values will become not only very possible, but also an imperative. If the imperative works, Klaus will overcome his split personality, and personalized sustainability can become daily routine.

References

Accenture and Havas (2014) The consumer study. From marketing to mattering. The UN global compact-Accenture CEO study on sustainability. http://www.havasmedia.co.uk/wp-content/uploads/Havas_Accenture-Consumer-Study_2014-low-res.pdf. Accessed 20 Apr 2017

Accenture Strategy (2015) Young fintechs capture use service innovation to capture the customer interface. Digital Startup Radar. https://vimeo.com/119332008

Accenture Strategy (2016) The digital emperor got no clothes. Are business leaders ready for a world of radical transparency? London. https://www.accenture.com/t00010101T000000__w__/gb-en/_acnmedia/PDF-34/Accenture-UK-Digital-Emperor-Video-Transcript.pdf

Accenture and United Nations Global Compact (2014) United Nations principles of responsible investments – the investor study: insights from PRI signatories. https://www.accenture.com/t20150523T042350__w__/us-en/_acnmedia/Accenture/Conversion-Assets/DotCom/Documents/Global/PDF/Industries_15/Accenture-Investor-Study-Insights-PRI-Signatories.pdf

Brookings (2016) Order from Chaos. What China's food safety challenges mean for consumers, regulators, and the global economy. https://www.brookings.edu/blog/order-from-chaos/2016/04/21/what-chinas-food-safety-challenges-mean-for-consumers-regulators-and-the-global-economy/. Accessed 28 Mar 2017

Comscore (2014) Major mobile milestones in May: apps now drive half of all time spent on digital. http://www.comscore.com/Insights/Blog/Major-Mobile-Milestones-in-May-Apps-Now-Drive-Half-of-All-Time-Spent-on-Digital. Accessed 6 Nov 2016

Fjord and Accenture (2015) Th era of living services. https://www.fjordnet.com/conversations/the-era-of-living-services/. Accessed 20 Apr 2017

Frankfurt Stock Exchange (2016) EON SE Share price history. http://en.boerse-frankfurt.de/stock/pricehistory/E.ON-share/ETR/29.6.2016_4.7.2016. Accessed 4 July 2016

Hofman R, van't Spijker A (2013) Patterns in data driven strategy. Five business model innovation patterns to create strategic value from data. BlinkPapers. http://www.blinklane.com/wp-content/uploads/2013/12/131212-BlinkPaper-DDS.pdf

Honest by (2016) Company website product. http://www.honestby.com/. Accessed 6 Nov 2016

Lacy P, Rutqvist J, Buddemeier P (2015) Waste to wealth (German version)

Metro Starfarm (2017) http://www.starfarmcc.com. Accessed 28 Mar 2017

Oral B (2016) Product information. http://www.oralb-blendamed.de/de-DE/oralb-smartseries-elektrische-zahnbuerste-mit-bluetooth. Accessed 2 Oct 2016

Renewable Energy Agency and TNS Emnid (2015) Acceptance poll: German citizens want more renewable energy (in German). https://www.unendlich-viel-energie.de/themen/akzeptanz-erneuerbarer/akzeptanz-umfrage/akzeptanzumfrage-erneuerbare-2015. Accessed 6 Nov 2016

Statista (2015) Green electricity statistics (in German). https://de.statista.com/themen/566/oekostrom/. Accessed 6 Nov 2016

The Economist (2015) The cheap, convenient cloud. http://www.economist.com/news/business/21648685-cloud-computing-prices-keep-falling-whole-it-business-will-change-cheap-convenient. Accessed 6 Nov 2016.

Is Digitalisation a Driver for Sustainability?

Carl-Otto Gensch, Siddharth Prakash, and Inga Hilbert

1 Introduction

Most processes in today's society are affected by digitalisation. Digital infrastructure is now an integral part of e-commerce, e-health, intelligent traffic control systems, energy production and transmission, smart appliances, insurance, and finance. Physical media are being replaced by digital structures (video-on-demand, IPTV and other platforms such as YouTube) for the archiving and storage of data, including photos and videos. Digitalisation is therefore having a great impact on our everyday lives. Whereas technical innovations and data security are often the focus of public debates, sustainability aspects are mostly neglected. But as the societal and environmental impacts of digital infrastructure grow, there is a need to refocus the ongoing debate. Apart from the direct environmental impacts of ICT products and infrastructure, the numerous impacts on business models, lifestyles and consumption patterns also need to be taken into account.

When it comes to the sustainability of digitalisation, two different approaches should be distinguished. "Green by IT" means making processes more efficient and sustainable by implementing IT systems. "Greening IT" aims to make the IT itself more sustainable. The main building blocks of a digitalisation strategy are data centres (storage, processing) and telecommunication networks (transmission).

This article focuses mainly on the direct impacts, and especially on resource consumption and environmental aspects. It therefore considers electricity consumption, related GHG emissions, and the product carbon footprint (PCF). In order to give a full overview, two different perspectives are included as well. In analysing the impact of digitalisation, this article considers both main approaches, and

C.-O. Gensch (✉) • S. Prakash • I. Hilbert
Oeko-Institut e.V.—Institute for Applied Ecology, Merzhauser Str. 173, 79100 Freiburg, Germany
e-mail: C.Gensch@oeko.de; S.Prakash@oeko.de; I.Hilbert@oeko.de

© Springer International Publishing AG 2017
T. Osburg, C. Lohrmann (eds.), *Sustainability in a Digital World*, CSR, Sustainability, Ethics & Governance, DOI 10.1007/978-3-319-54603-2_10

discusses the results of top-down and bottom-up study approaches. Finally, the article summarises the results from the different studies and concludes with recommended future fields of action.

2 Top-Down: An Unprecedented Increase in the Electricity Consumption of Data Centres and Telecommunication Networks in Europe

In a study conducted by Oeko-Institut and Technical University of Berlin on behalf of the European Commission, the electricity consumption of data centres was forecast to increase by almost 35% from 52 TWh in 2011 to 70 TWh in 2020 (Prakash et al. 2014[1]). At the same time, the electricity consumption of telecommunication networks was forecast to increase by a massive 150% from 20 TWh in 2011 to 50 TWh in 2020. The biggest growth is anticipated for the mobile networks, due to immense growth in mobile data traffic—by a factor of 30[2]—caused by the more intensive use of mobile internet services. Main drivers of this trend are new cloud-based services (storage in the cloud, software as a service, apps), more time spent online (at home and mobile) with new end-user devices, increased use of videos—uploads and downloads—(YouTube, online streaming, video on demand, IPTV in high definition), social media with frequent status updates, photo uploads, and so on. These are enabled by more capable mobile networks (LTE technology) as well as an increasing number of mobile devices with significant computing power (smartphones, tablets). Recent data from market analysts IDC shows that 1.301 billion smartphones[3] and 229.6 million tablets[4] were sold globally in 2014.

Compared to the enormous increase in the mobile traffic, landline traffic is set to grow by a relatively moderate factor of 3.[5] According to Prakash et al. (2014), high growth in the electricity consumption of data centres is anticipated due to increased data traffic per subscriber. This is reflected in the fixed data traffic forecast from

---

[1]Prakash, S.; Baron, Y.; Liu, R.; Proske, M. & Schloesser, A. (2014). Study on the practical application of the new framework methodology for measuring the Environmental impact of ICT—cost/benefit analysis (SMART 2012/0064). Oeko-Institut e.V. and TU Berlin for the EU Commission, DG Communications, Networks, Content & Technology, Brussels.

[2]Cisco Visual Networking Index (VNI) Forecast Widget, used data was assessed online in June 2013; http://www.ciscovni.com/forecast-widget/advanced.html.

[3]In a near tie, Apple closes the gap on Samsung in the fourth quarter as worldwide smartphone shipments top 1.3 billion for 2014. Press release International Data Corporation (IDC). Available at http://www.idc.com/getdoc.jsp?containerId=prUS25407215, last accessed 25 Jan 2016.

[4]Worldwide tablet shipments experience first year-over-year decline in the fourth quarter while full shipments show modest growth. Press release International Data Corporation (IDC). Available at http://www.idc.com/getdoc.jsp?containerId=prUS25409815, last accessed 25 Jan 2016.

[5]Cisco Visual Networking Index (VNI) Forecast Widget, used data was assessed online in June 2013; http://www.ciscovni.com/forecast-widget/advanced.html.

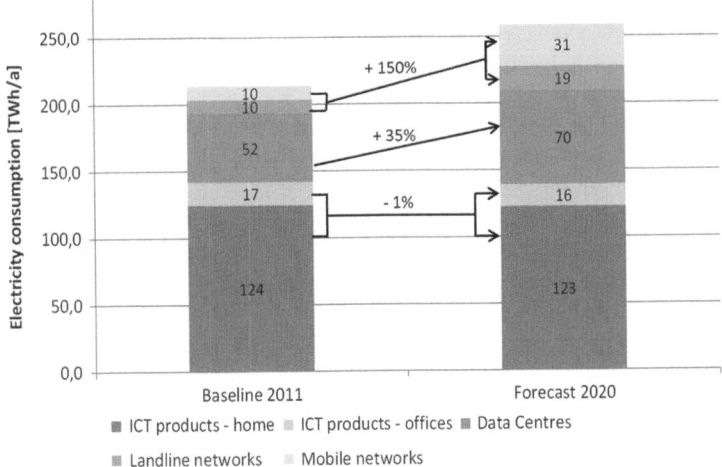

Fig. 1 Comparison of the ICT-related electricity consumption in the EU-27 in 2011 & 2020 (excluding ICT manufacturing)

Cisco, of about 22 EB[6] per month in 2016. The demand for cloud services like SaaS[7] or PaaS[8] is thus expected to grow dramatically in the near future. Such a trend will influence not only the server market, but also the storage market which has already seen very strong growth in recent years. The resultant increase in internet and cloud service usage, and with that, the electricity consumption of data centres and networks, can therefore be seen as one of the logical consequences of increasing digitalisation in our daily lives. In any case, the collective share of data centres and telecommunication networks in total ICT-related electricity consumption in the EU-27 is expected to increase from 33% in 2011 to about 46% in 2020. In other words, data centres and telecommunication networks will account for around 3.8% of the total electricity consumption of the EU-27 in 2020, compared to 2.6% in 2011 (Prakash et al. 2014).

As far as the total electricity consumption of ICT in the EU-27 is concerned, the modelling and calculations carried out by Prakash et al. (2014), show that the total ICT-related electricity consumption (excluding manufacturing) in the EU-27 is expected to increase from 214 TWh in 2011 to 259 TWh in 2020 (see Fig. 1).

Thus, the share of ICT-related electricity consumption, in the use phase, in EU-27 is expected to increase from 7.7% in 2011 to 8.1% in 2020 (Prakash et al. 2014). The 214 TWh electricity consumed by ICT in 2011, correspond to the yearly consumption of more than 61 million households.[9]

---

[6]Exabyte (EB) = 10^{18} Bytes.

[7]Software as a service.

[8]Platform as a service.

[9]Based on the average consumption of 3500 kWh per household.

As shown in Fig. 1, use of ICT products (home and office, including televisions) still has the largest share with approximately 66% of total ICT-related electricity consumption in 2011. In absolute terms, the electricity consumption of ICT products in 2011 in the EU-27 was 142 TWh/a. TVs had the biggest share of all products, followed by desktop PCs, games consoles and home gateways. Although the stock of those end-user ICT products will have the strongest growth among all categories in the coming years, a decrease in electricity consumption to 139 TWh/a in 2020 can be expected. This is attributable to the broader use of mobile products, decreased use of DECT phones in offices, and substantial energy efficiency improvements for the use phase of TVs, desktop PCs, computer displays and notebooks.

As far as the electricity consumption of ICT at the global level is concerned, a study by Corcoran and Andrae (2013)[10] estimated that global electricity consumption relating to ICT (including manufacturing) will increase from 1817 TWh in 2012 to 1982 TWh (best-case scenario) or as much as 3422 TWh (worst-case scenario) in 2017. Converting the electricity consumption figures into GHG emissions—using the carbon emission factor applied in the EuP EcoReport tool developed for the EuP Directive 32/2005/EC, i.e. 0.4582 kg CO_2 e/kWh)—the global ICT-related greenhouse gas emissions would be expected to increase from 0.75 Gt in 2012 to 1.05 Gt in 2017 (Prakash et al. 2014).

3 Bottom-Up: Product Carbon Footprint of Online Storage Services

Over the past decades, ICT products have found their way into private households on a large scale. The expansion of these devices is one of the reasons that absolute power consumption has remained the same despite efficiency gains in individual household appliances.

Alongside these ICT end-products in private households or offices, the associated infrastructures must also be taken into account. Besides telecommunication networks, data centres are also relevant. However, these can vary greatly in terms of performance and technical equipment. As regards the Blue Angel eco-label for energy-efficient data centre operation services, energy usage effectiveness (EUE) is used to measure energy efficiency in data centres. EUE describes the ratio between the entire data centre's annual demand for energy (including power supply, cooling, lighting, etc.) and the energy requirements of the actual ICT system (server, storage, network, etc.) The greater the EUE value, the less efficient the data centre. An energy-efficient data centre approaches a ratio of 1. To meet the eco-label

---

[10]Corcoran, Peter & Andrae, Anders (2013); Emerging Trends in Electricity Consumption for Consumer ICT; http://vmserver14.nuigalway.ie/xmlui/handle/10379/3563.

requirements, the EUE value should not exceed a ratio of between 1.4 (new data centres) and 1.8 (older data centres), depending on when they started operation. In addition, requirements are imposed in relation to the individual components as well as energy management. The operator has to subject the data centre to a continuous improvement process aimed at optimising the efficiency of energy usage. This should be reflected in declining EUE values. Use of the eco-label requires an external audit to be conducted, as well as regular reporting to RAL, the awarding authority.

In a recently completed research project (Gensch et al. 2016)[11], an approach was chosen which is opposed to the previous one. Specifically, the project looked at whether it is possible to derive climate-relevant criteria by focusing on typical IT services offered by data centres as intermediaries. So the idea is to examine whether climate-relevant savings potentials can be opened up by IT services-related criteria that go beyond the criteria relating to end-products and data centre operations. The investigation therefore (proportionately) includes all components which are required for providing and using IT services. These are

- ICT devices,
- the required networks and
- data centers.

Originally, several IT services provided by data centres were to be evaluated together—for example the five most important ones—based on this hierarchical model, as a second case study in the project. But it turned out that analysing the horizontal and vertical service dependencies throughout and between the described levels would take far too long given the time available for the project. Therefore it was decided to limit the examination to a typical service. And so the "online storage" service was selected.

The major function of online storage services is to exchange files between computers, tablets and smartphones and to ensure their synchronisation. Due to the increasing popularity of mobile devices in particular, online storage services represent a rapidly growing market segment. However, this segment is also subject to strong competition and sharp decreases in prices.

Furthermore, groups like Microsoft are continuing to develop the cloud infra-structure, since it can be assumed that demand will continue to grow strongly. The product carbon footprint (PCF) calculation carried out in this project was aimed at identifying the main factors influencing these IT services. Analysing contributions to the PCF along the life-cycle of the required system components should yield

---

[11]Gensch, C.-O.; Liu, R.; Prieß, R.; Stratmann, B.; Teufel, J.; Product Carbon Footprint: Möglichkeiten zur methodischen Integration in ein bestehendes Typ-1 Umweltzeichen (Blauer Engel) unter besonderer Berücksichtigung des Kommunikationsaspektes und Begleitung des Normungsprozesses [Product Carbon Footprint: Options for methodological integration into an existing type 1 eco-label (Blue Angel) with special emphasis on communication aspects and on monitoring of the standardization process]. Study on behalf of German Federal Environment Agency (UBA), Dessau. Freiburg 2016.

clues as to where potential hotspots are located. On this basis, the next step would be an analysis to determine whether the identified hotspots and influencing parameters could serve as starting points for improvements, and whether climate-related criteria for awarding an eco-label can be derived from this. The case study was prepared in close cooperation and consultation with the Chair of Information and Communication Management at TU Berlin. Several existing studies on the use phase of online versus offline storage options were used as references. Building on these investigations, the following amendments were made for the present study:

- In addition to the investigated use phase, the production phase of the IT components used (or taken advantage of by using the service provided by data centres) was also included in the scope, since it can be assumed on the basis of other studies that a considerable share of the carbon footprint is due to this phase.
- Moreover, several sensitivity analyses were performed to gain better estimates regarding the significance of assumptions relating to the conditions of use when determining the functional unit, for example, and regarding the importance of influencing parameters.
- Finally, the assumptions made and results calculated in this study were situated in an overall context with other studies.

Essential data on online storage has been collected by the Technical University of Berlin in cooperation with a large internet service provider (ISP). This ISP offers its customers web-hosting, domain and e-mail services, as well as server-hosting and cloud services. In the work carried out by TU Berlin, online storage services have been investigated, and also online and offline storage have been compared with one another. For the offline storage option, the use of a network-attached storage (NAS) system exclusively operating in home networking was assumed. In the present study, and in a manner similar to that employed by TU Berlin, this offline solution was also taken into consideration, resulting in greater transparency concerning the relevance of influencing parameters. However, it was neither intended nor possible to produce a comparison of online and offline data storage systems. This would require a more thorough analysis based on representative usage data, which is beyond the scope of this study. Furthermore, the considered systems were not substantially functionally equivalent.

The analysis shows that, under the methodological specifications laid down here, the annual use of online storage with a daily upload volume of 1.2 GB and a download volume of 1.0 GB is associated with 58 kg CO_2e per year, and that of offline storage with 98 kg CO_2e per year. This result, however, depends very much on the assumed conditions of use. Consequently, reliable conclusions made on a comparative basis, such as "according to climate protection considerations, online storage is superior to offline storage", cannot be made. The following results come from an analysis of the respective contributions:

- Making up approximately 80% of overall GHG emissions, the use phase is the dominant stage in each of the two alternatives.

- With the online storage option, two-thirds of GHG emissions can be attributed to the user of this service, while one-third stems from the data centre service provider. The internet as a "data transfer medium", however, only accounts for about 4% of GHG emissions.
- While the greenhouse gas emissions produced by offline storage depend almost entirely on the NAS device used (around 80% being attributable to the use and 20% to the manufacturing phase), the contributions in the online storage option are unevenly distributed. The fact that using a laptop at work accounts for approximately one-third of the overall emissions is striking, and demands explanation. This share is attributable to the requirement that the laptop must remain switched on during data uploads and downloads, and hence the corresponding energy demand for this function takes effect. Due to the device's high utilisation time, its production—according to the allocation model—also accounts for a proportion of 8% of the overall emissions. The same applies to the share of the LAN net attributable to the user (18.4% of GHG emissions). Other relevant shares are to be allocated to the level of the data centre: here, too, it is interesting to note that the subsystems relevant for the free movement of data— namely gateway and LAN with 9.6% and 1.4% respectively—together account for roughly the same percentage as the server with the hard disk systems used for data storage (12.1%).

Basically, the methodological approach that is tested by way of example here, in the case study for storage services—i.e. to look at an IT service as an integrated system to be investigated over its entire life-cycle—results in a presentation of results and a view of influencing parameters derived from them which cannot be deduced from an approach and analysis that only takes into account data centres and ICT devices used by users. Furthermore, it becomes clear that the share of these services in the GHG emissions produced by a private household in the area of ICT equipment is by no means negligible, but can be assumed comparable to the level of the use of television sets, for example.

The results for the online storage option show that relevant shares in the GHG emissions associated with this service are attributable both to the required use of IT devices in the user's household, and to nets and data centres. The comparison for guidance purposes with the offline storage alternative—i.e. using an NAS device in home networking—which was undertaken for a proper understanding of the results, shows the high relevance of the conditions of use in influencing the results. Accordingly, on the basis of this case study, no unequivocal advantages or disadvantages can be attributed to either storage option in respect of their relevance to climate protection. In an extensive use scenario, online storage tends to be environmentally superior to an offline solution. Conversely, in the case of heavy use, offline use will be associated with lower GHG emissions than online storage. Given that only one service provider and one example configuration for an offline solution could be investigated in this case study, no reliable "break-even point" can be derived. Finally, it should be noted again here that the two alternatives only partly exhibit the same functionalities, and therefore, in that respect, can only to a limited extent be considered as equivalent and comparable.

The high level of differentiation in the use of ICT-related services is likely to be one of the major difficulties encountered when comparing different providers and their offerings with regard to their respective GHG emissions. For example, the online storage services provider who was considered in this study alone offers five different products that vary in storage capacity and the number of data transfer accesses. The result of this high level of specialisation, together with the substantial need for data on performance parameters from the data centres that are used to provide the services, is to hinder comparability, a ranking, or the setting of minimum standards in order to derive criteria as a basis for drafting fundamental award principles.

It must also be remembered that fundamental award principles for the Blue Angel eco-label already exist for data centers as well as for the ICT devices which are relevant here. The criteria set out in these fundamental award principles already address the crucial influencing parameters that are relevant for online storage as well as for the offline storage option. Any further-reaching conclusions in terms of additional criteria can be drawn only to a very limited extent, based on the analysis of storage services undertaken in this study:

- As described above, in online storage, a higher transmission rate for internet data would reduce the utilisation period of the IT infrastructure for the user, and thus the energy requirements of these components. Whether and to what extent higher transmission rates increase the energy and hardware requirements in the network, and how this relates to the energy savings for the user, cannot be ascertained from the available information. It must also be remembered that transferring large amounts of data, given the average transfer rates prevailing at the moment, has a limiting effect for the user. Therefore it cannot be ruled out that an increase in transmission rates will also entail rebound effects.

- As regards the offline storage option and the NAS devices used for this purpose, devices with a smaller number of hard drives might reduce GHG emissions. But the question as to what extent failure security and protection against data loss would be affected in this case can only be answered with difficulty. On the other hand, the joint usage of NAS devices in households could help reduce energy demand and GHG emissions on a pro-rata basis. In the context of further development of the fundamental award principles, consideration might be given to whether devices available on the market differ from each other in terms of software installation and administration of multiple users, and whether joint usage of NAS devices might be encouraged through an additional criterion designed for this purpose.

4 Bottom-Up: E-Books

E-books are another example of an ICT-based service that is becoming increasingly relevant. Looking at total sales and the number of published titles, the market for books in Germany is still largely dominated by print books. The German book market was worth 9536€ million in 2013 and 9322€ million in 2014. E-books had a revenue market share of 3.9% in 2013 and 4.3% in 2014 (Börsenverein des Deutschen Buchhandels 2016). By contrast, they had a share in the U.S. book market of more than 30% (Graef 2016).[12] Crucial factors explaining this difference could be the established structure of bookshops in Germany, which is supported by legal price-fixing for books. Then again, new business models for the distribution of e-books—like Amazon Kindle Unlimited, a flat-rate model—were introduced in Germany 2014, some years after they appeared in the United States. In this light, and taking into account consumer surveys on future buying intentions (Berg 2015),[13] there is a strong suggestion that the market for e-books in Germany will see considerable growth in the years ahead.

One frequently made argument in favour of e-books is that paper (and the wood to produce paper fibres) along with other materials needed to produce printed books could be saved. Therefore, e-books could be understood as a strategy of dematerialisation through the replacement of physical products with virtual goods. It is argued that the replacement of physical products by digitalisation promotes environmental sustainability. However, there is no doubt that this advantage has to be weighed against specific properties of e-books, especially considering raw materials and the energy needed for production, the use-phase and end-of-life of devices which are needed to read the electronic files.

With the help of a life-cycle assessment (LCA)—an integrated method developed and standardised back in the 1990s—it is possible to compare the environmental impacts of print books vs. electronic books. There are several LCA studies available that compare printed media with electronic alternatives. A handful of these studies specifically focus on e-books to determine their environmental impact and compare them with paper books. In a recent research project,[14] a meta-review of available LCA studies relating to e-books and/or paper books was conducted. The results of these studies are not easy to assess, given that paper books and e-books differ considerably in several significant parameters. With conventional books, the decision to produce a hardcover or paperback, as well as the choice of paper (recycling vs. virgin fibres, grammage), has a major effect on the environmental impact. E-books might be read on specific e-reader devices, but also on other electronic devices like PCs with LCD displays, tablet PCs, laptops or

---

[12]Graef, Ralph Oliver (2016); Recht des E-Books und des Electronic Publishing. C.H. Beck, München.

[13]Berg, A.; Studie zur Nutzung von E-Books. Vortrag Pressekonferenz Bitkom, Berlin 06. Oktober 2015.

[14]http://www.trafo-3-0.de/index.php?id=2, accessed 08/04/2016.

notebooks and smartphones. E-readers with their specific e-ink displays offer some advantages compared to these devices, as they share many of the characteristics of paper (comfortable to read, wider viewing angle than LCD displays and better contrast ratio). In addition to these usage properties, e-readers are beneficial from an environmental perspective due to their low power consumption in the use phase compared to other devices used to display e-books. In the following, we will discuss some main results from Moberg et al. (2011),[15] as this study considers a broad range of different environmental impact categories and offers a integral discussion of scope and assumptions. With regard to e-books, the authors clearly identify the production of the e-book reader as the main contribution to the environmental impact. Due to the use of e-ink displays, the energy consumption for reading is significantly reduced. Accordingly, the reading time, which has significant influence on the result for other devices, is no longer an issue. Moberg et al. compare e-books to paper books along a set of 11 different impact categories. Assuming a base scenario that implies a total reading of 17,000 pages (corresponding to 48 books), the e-book was preferable to the hardback paper book studied here in terms of resources used, global warming, energy, eutrophication, human toxicity, marine aquatic ecotoxicity and terrestrial ecotoxicity. In contrast, the paper book was preferable in terms of acidification, ozone depletion, freshwater aquatic ecotoxicity and photochemical ozone creation. It is shown that the environmental impact of an e-book is dependent on the total use of the e-book reader. If the electronic device is used for very few books, paper books were preferable from an environmental perspective. For several impact categories (climate change, abiotic depletion, eutrophication, human toxicity, marine aquatic ecotoxicity and terrestrial ecotoxicity), Moberg et al. identified a break-even of around 30 books (see Fig. 2, "cumulative energy demand"). In this figure, the x-axis shows total number of books read and the y-axis the cumulative energy demand per book read. If paper books are assumed to be read twice (shared use by two readers, which would not be easily possible in the case of e-books) the break-even for these impact categories increases to around 60–70 books. Given these findings, the authors conclude that there is no single answer as to which book is better from an environmental perspective. The comparison depends to a considerable extent on parameters that vary for each book and user. As an improvement option, an e-book reader should be used by frequent readers, and, if possible, for different purposes such as reading books, newspapers, journals and other documents, thus lowering the impact per functional unit. Furthermore the life-time of the e-reader should be prolonged as far as possible.

Generally these conclusions have to be discussed taking into account the following aspects:

• Due to data gaps no data was available for the e-ink screen.

[15]Moberg, Å.; Borggren, C.; Finnveden, G. (2011): Books from an environmental perspective—Part 2: e-books as an alternative to paper books. Int J Life Cycle Assess (2011) 16: 238–246.

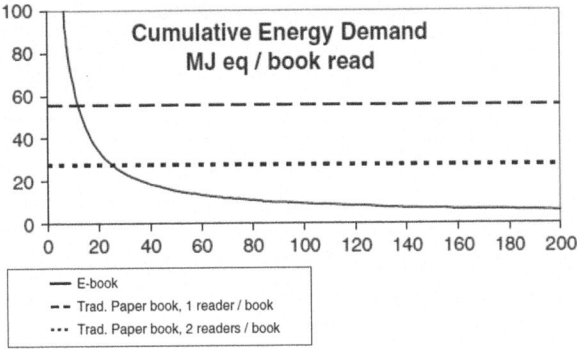

Fig. 2 Cumulative energy demand per e-book as a function of total amount of e-books read on an e-book reader (Moberg et al. 2011)

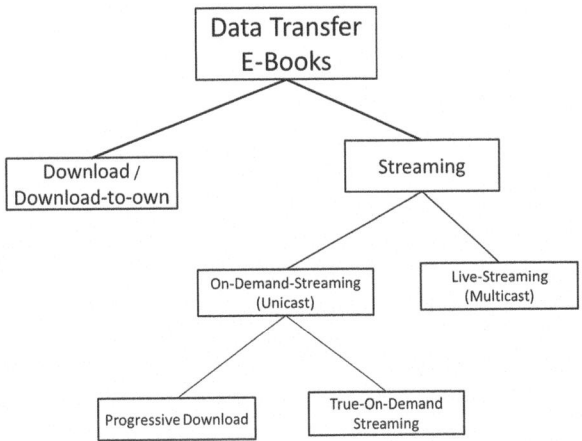

Fig. 3 Different options for the delivery of e-books

- The study considers only the use of an e-reader. According to a current consumer survey for Germany in 2014 (Berg 2015), readers mainly use notebooks/laptops (41% of 577 users), followed by smartphones (38%), e-readers (34%), PCs (21%) and tablet PCs (20%). Using devices other than an e-reader, energy consumption in the use phase and especially the reading time would be crucial as these are the main parameters with regard to the energy consumption of the system.
- The study assumes that data transfer is done exclusively via download, using a PC to access the internet. According to the business-model trend over recent years and the results of the consumer survey mentioned above, data transfer today will be done mainly by on-demand streaming (see Fig. 3). However, continuous access to the cloud augments the energy demand of the device as well as the energy demand of the network and data centre.

It is certain that all of these aspects would increase the energy consumption and accordingly the environmental impact of e-books.

Generally, digitalisation of book-making affects not only the way in which information between authors and reader is delivered. Rather, the way that books are written as well as the way of reading is changing, too. In order to sustain cultural diversity, publishing companies have to develop new services and redefine their role. Thus it is evident that the ongoing transition to e-books also affects cultural aspects. The world of conventional paper books in Germany currently comprises about 5000 bookshops as well as 8200 public libraries. In contrast, e-books are distributed by less than five providers. Furthermore, these providers exclusively collect readers' related data in a depth that far exceeds former statistics on reading behaviour. Some initiatives argue that society should not leave our literary cultural heritage as well as data on readers' behaviour in the hands of a few multinationals. Instead, non-profits should ensure a neutral infrastructure for the literary scene (see for example log.os).

5 Main Conclusions

The preliminary analysis illustrates the impact of digitalisation on the environment. In the EU, the overall share of ICT-related energy consumption is expected to increase further over the next few years. Especially the energy consumption of data centres and telecommunications networks is expected to grow significantly by 2020. At the same time, there is a lack of policy measures for regulating resource and energy consumption, and GHG emissions of data centres and telecommunication networks. The lack of policy measures can be attributed partly to a lack of publicly available data, e.g. on the energy consumption and GHG emissions of data centres and telecommunication networks. While the required data is not always available yet, the necessary methods for impact assessments do already exist. However, methods like life-cycle costing, greenhouse gas protocols and life-cycle assessment can already be used to estimate the environmental impact of ICT. At the same time, there are generally valid options that can diminish the environmental impact (such as using electricity from renewable energy sources).

One main problem is the absence of critical investigations into the ongoing development trends. The environmental and societal impact of technological innovations and corresponding new business models are not analysed in advance. Consequently, such new services are mostly not designed from a sustainability point of view. Although it is not currently known whether or not it is more sustainable to store data and software externally in clouds, the growth of these services is unrestricted. There is a lack of understanding among political decision-makers about the possible sustainability impacts of the digitalisation of our daily lives. While the focus has been on highlighting the positive effects of digitalisation, the possible rebound effects and unexpected consequences leading to risks have so far been underestimated. This can be seen for instance in research trends and funding on the topic of cyber-physical systems and the Internet of Things (also called *Industry 4.0* in Germany), which can be termed largely technocratic in their approach. Only a few supported projects specifically address sustainability or aim to reduce negative environmental impacts.

While there is an ongoing debate on data security and the control of our personal data in the context of digitalisation—largely in the wake of recent scandals, such as those involving national security agencies—the social and environmental dimensions of a digitalisation strategy have not yet been adequately addressed. Therefore, digitalisation needs a political framework to ensure that its development takes sustainable development (goals) into consideration. The multi-level perspective (MLP) could be used to develop this framework. To describe fundamental changes such as digitalisation, the MLP takes into account different levels of a changing system. A comprehensive analysis of the whole system and its interdependences allows us to develop a framework which goes beyond regulatory measures and aims to provide fitting solutions.

References

Berg, A (2015) Studie zur Nutzung von E-Books. Vortrag Pressekonferenz Bitkom, Berlin

Börsenverein des Deutschen Buchhandels (2016) Key figures about the e-book market in Germany from 2013 to 2015, quoted by statista.com. Available via http://www.statista.com/statistics/385743/e-bookmarketkey-figures-germany/. Accessed 8 Apr 2016

Cisco Visual Networking Index (VNI) Forecast widget, used data was assessed online in June 2013. http://www.ciscovni.com/forecast-widget/advanced.html

Corcoran P, Andrae A (2013) Emerging trends in electricity consumption for consumer ICT. http://vmserver14.nuigalway.ie/xmlui/handle/10379/3563

Gensch C-O, Liu R, Prieß R, Stratmann B, Teufel J (2016) Product carbon footprint: Möglichkeiten zur methodischen Integration in ein bestehendes Typ-1 Umweltzeichen (Blauer Engel) unter besonderer Berücksichtigung des Kommunikationsaspektes und Begleitung des Normungsprozesses [Product carbon footprint: options for methodological integration into an existing type 1 eco-label (Blue Angel) with special emphasis on communication aspects and on monitoring of the standardization process]. Study on behalf of German Federal Environment Agency (UBA), Dessau, Freiburg. http://www.trafo-3-0.de/index.php?id=2. Accessed 8 Apr 2016

Graef RO (2016) Recht des E-Books und des Electronic Publishing. C.H. Beck, Münche

In a near tie, Apple closes the gap on Samsung in the fourth quarter as worldwide smartphone shipments top 1.3 billion for 2014. Press release International Data Corporation (IDC). Available at http://www.idc.com/getdoc.jsp?containerId=prUS25407215. Accessed 25 Jan 2016

Moberg Å, Borggren C, Finnveden G (2011) Books from an environmental perspective—part 2: e-books as an alternative to paper books. Int J Life Cycle Assess 16:238–246

Prakash S, Baron Y, Liu R, Proske M, Schloesser A (2014) Study on the practical application of the new framework methodology for measuring the environmental impact of ICT—cost/benefit analysis (SMART 2012/0064). Oeko-Institut e.V. and TU Berlin for the EU Commission, DG Communications, Networks, Content & Technology, Brussels

Worldwide tablet shipments experience first year-over-year decline in the fourth quarter while full shipments show modest growth. Press release International Data Corporation (IDC). Available at http://www.idc.com/getdoc.jsp?containerId=prUS25409815. Accessed 25 Jan 2016

Sustainable Digital Business: Crucial Success Factor for Small and Medium-Sized Enterprises and Start-Ups

Franz Wenzel

1 Digitization and Sustainability

How can Small and Medium-Sized Enterprises (SME) and Start-Ups address digitization in a sustainable way and how can such companies profit from the crucial success factor of Sustainable Digital Business?

Digitization as Megatrend

Digitization is one of the most extensive Megatrends of our times.

Megatrends (Naisbitt, 1982), such as *globalization, urbanization, digitization, networking* and *automation* (and others) shape this world faster and bring more change than ever before in human history (cp. Fig. 1). Never before more people in more places have been affected by change that fast and intense. Besides we can consider trend effects to speed up over time, especially when Megatrends mix or combine. Simplified, new Megatrends evolve out of the paths of old trends or throughout the combination of trends.

Slipstreaming these and other big trends, companies try to foster their strategic and/or financial goals (cp. Fig. 2). Digitization (as an example of the Megatrends) might easily bring up new products, open new markets, create new ways of communication or offer economies in costs.

Figure 2 also illustrates the position of the different company structures and sizes in the field of Megatrends. It's often SMEs/Start-Ups that use the big trends as first movers or find paths to applications or customers. Non Governmental Organizations (NGO) or Non Profit Organizations (NPO) often also lead development, whether they use a trend to foster their aims or they position themselves towards a certain development. Big Companies (BC) tend to follow in a short distance, often

F. Wenzel (✉)
Dipl.-Kfm. Franz Wenzel, Digital Frontier Academy, Kreuzstraße 14, D-85049 Ingolstadt, Germany
e-mail: franz.wenzel@cooperative-innovation.com

© Springer International Publishing AG 2017
T. Osburg, C. Lohrmann (eds.), *Sustainability in a Digital World*, CSR, Sustainability, Ethics & Governance, DOI 10.1007/978-3-319-54603-2_11

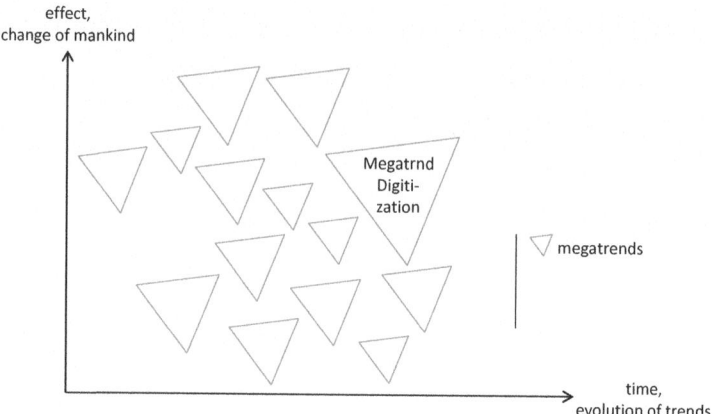

Fig. 1 Megatrends

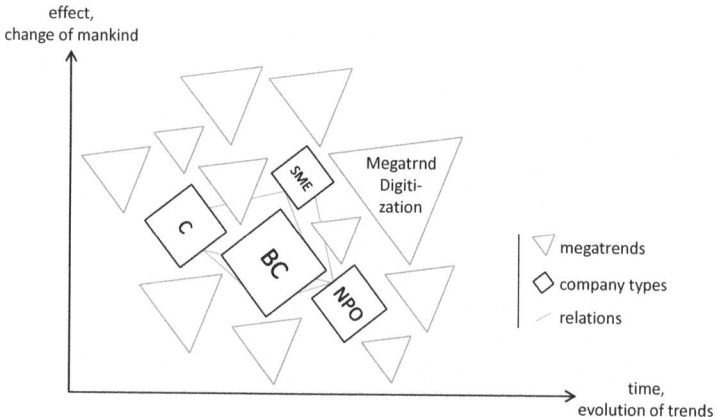

Fig. 2 Megatrends, companies and relations

using the innovations provided or the path prepared by SMEs/Start-Ups or NPOs. In most cases established companies of average size follow at a certain distance, as they are not sure If they can or want to follow a trend at an early stage of the trends' development. The different companies are interlinked and related in various ways.

A short look at the role of big companies in the field of Megatrends:

As big companies accumulate various kinds of resources and do their business in a broader scale, they tend to combine/link resources and bridge technological and cultural gaps. Doing so, they speed up the development and the spreading of trends and Megatrends. This is true not only at the good ends of innovation but also on all downsides any development might bring.

As Megatrends cannot be compared with and are not the same as fashion, such big trends are irrefutable. They do bring change to all mankind and they do not vanish in thin air after one or two seasons. One can neither positively nor negatively decide to accept a Megatrend as it is already happening and it is running system wide.

A short look at open societies:
Open societies—the political, socio-cultural and legal system, including the relevant approaches towards religious, international and socio-intercultural exchange—also foster big trends as there is a very intense system of a special form of competition between the citizens of such societies in the sense of reaching higher respectively more individual levels of self-fulfillment and satisfaction. This is usually done by implementing new ideas and trends (at least in significant parts of these societies) very fast and with long-range effects into education, lifestyle and culture.

A short look at societal tensions:
Societal tensions occur within such societies and between different societies when one part follows a trend and the other part is excluded or excludes themselves from the trend or the ability to follow. This effect is more intense when the level of change is more abrupt or in its result very different for different parts of society. Digitization might be a good example on this effect as e.g. some people find new jobs or even private/personal fulfillment in its usage where other people are not sure how or if they should open themselves to the respective developments.

Some Brief But Important Words on Sustainability

Sustainability defines the target function of life: use as much resources as you need plus as much as you want (do not want too much, you can only carry a certain amount of weight), given the needs and wishes of others (do not forget them, they most likely are stakeholders of your life or work) today and tomorrow, restricted by limited resources respectively a limited system capacity (in the ecological dimension our world: i.e. earth, oceans and atmosphere) to handle emissions and waste (again in the ecological dimension: our world will fail catastrophically if we do not accept the latter factor). Limited system capacity also applies to the capacity a society can offer (e.g. the number of refugees, the number of unskilled workers, etc.) and to financial markets (e.g. overheating, availability, profitability, etc.) The System Capacity Horizon (SCH) tries to illustrate this important insight (cp. Fig. 3).

To the economic scientist the introduction of need and want, given any form of scarcity (of resources), implies a market. Thus the whole world undoubtedly can be understood as a system of markets.

A short look the concept of market used in this text:
This article neither uses market as such as a solution concept, nor implies market any form of economic system (such as market economy, capitalism, etc.). Market implies that any form of implicit (in exchange for the love of children or in exchange for respect) or explicit (in exchange for money or manpower) exchange is made to deal with scarcity (of resources, of time, of space, of innovation, etc.).

A more profound understanding of sustainability needs insights in the concepts' manifold dimensions and connections to market solutions, scarcity and competition

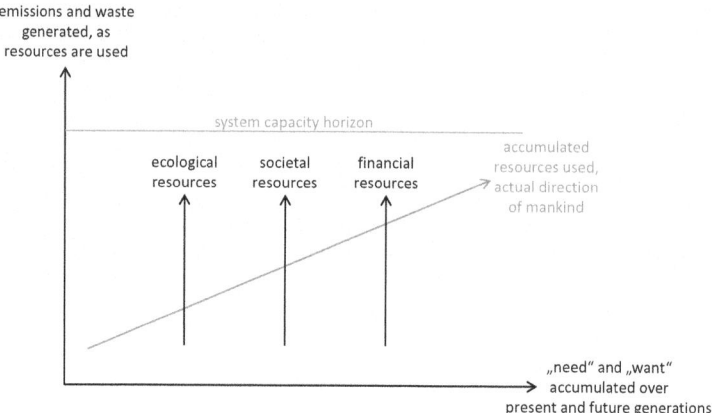

Fig. 3 Sustainability, accumulation of resources used towards system capacity horizon

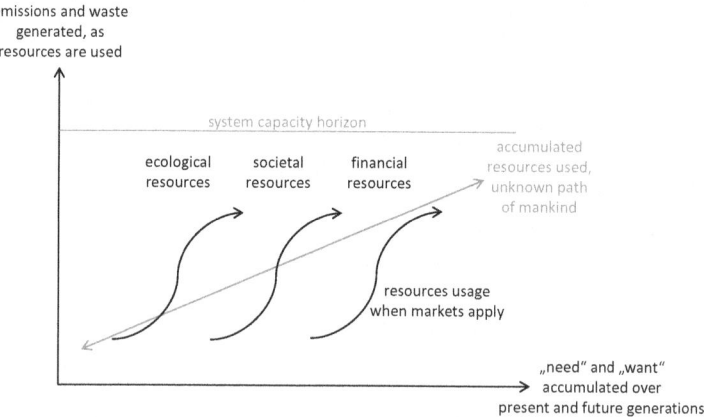

Fig. 4 Sustainability, resources and markets

(Smith 1776). And Sustainability is not only about ecology. It holds a financial dimension as well and its social realm might be the most fascinating part of the whole concept (cp. Fig. 4).

Digitization Meets Sustainability

In Fig. 5, Digitization (as an example of a Megatrend) is heading towards an (idealistic) direction of mankind. In the figure the direction could be considered "forward and up" if some technocratic mind is processing the figure. From a more holistic perspective the translation for the direction of mankind could be "sustainability" or a "well balanced state of true satisfaction for all mankind today and tomorrow" (see more on this in the final chapter of this article).

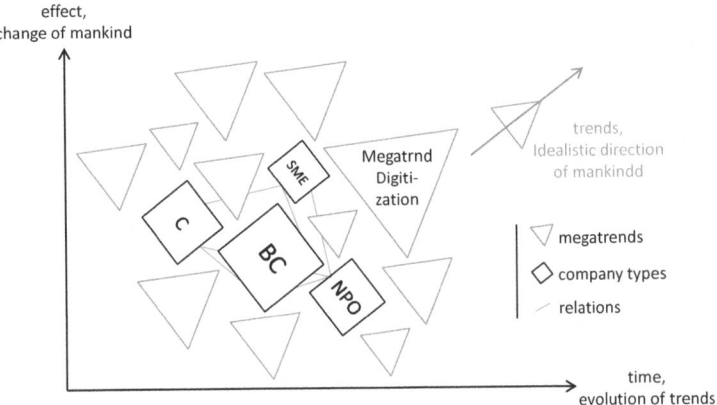

effect,
change of mankind

trends,
Idealistic direction
of mankindd

megatrends

company types

relations

time,
evolution of trends

Fig. 5 Megatrends, idealistic direction

Figure 6 introduces the market, scarcity and competition, shown as a dimension arrow in two directions. If, for example, lots of ecology is used (i.e. pollution rises), the costs for ecology might shoot up, too (costs for better filters, penalty payments, less satisfied customers). If an entrepreneur then wants to find financial balance, he/she might accept to use less ecology (i.e. reduce pollution) to reduce costs. Given the System Capacity Horizon and under conditions of a market, sustainability seeks a balance between the resources used respectively waste/pollution produced and the sum of need plus want.

It is important to understand that waste and pollution are used from the ecological perspective synonymously for the societal and financial dimension. Of course it is also true that the use of societal and financial resources also produces byproducts, so that the terms "waste" and "pollution" are used similarly and in a technical form. "Unemployment" as an example for the societal market or "instability" in the financial domain could be the terms to be used on these markets.

Sustainability—in an entrepreneurial economy—needs creative approaches. Doing more with less and doing things better with less ecological, financial and societal impact on the downside and with more ecology, with better financial stability and result and with a balanced societal effect on the sunny side—and doing so towards a stable equilibrium—needs ongoing innovative processes in the influence sphere of the users (how they use and dispose products and how they start and finish services) and the producers (what they produce or do and how it is made respectively recycled).

A short look at the roles of users and producers:
Surely there is no clear separation between the roles of users and producers any more. And this—of course—has various effects on creativity and innovation. The effect is not only a result of the fact that producers often are users of previously produced products (the suppliers provide goods for the production processes in the supply chain) in more intense or complex forms. It is more important that new roles apply as users can now be producers easier than in former times. Although products are increasingly complex, production methods as well as pre-products generally can be understood and used easier, better and

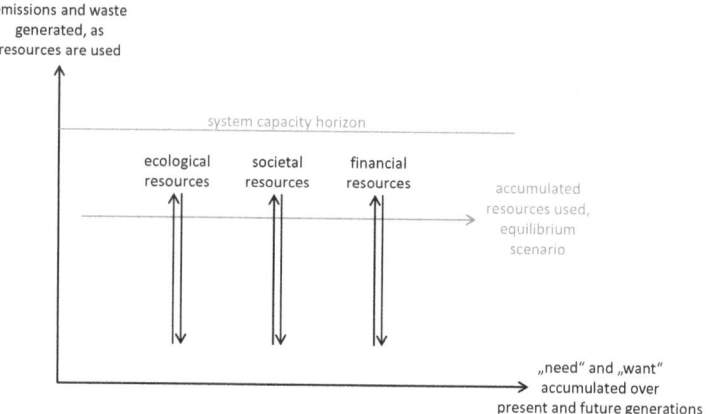

Fig. 6 Sustainability, equilibrium in the system, usage of resources and trends

faster as in former times. The complexity of things is reduced by a clearer separation of work and more flexible products and approaches. Same as 3D-Printing (flexible production method) is nowadays available to a broad public, do-it-yourselfers can easily apply special products e.g. in home renewals or creative approaches (flexible pre-products).

Of course innovation follows the paths of Megatrends, as such trends are running system wide. Hence business done the sustainable way might bring up digitization as promising innovative solution (and vice versa). Digital publications at first sight do not need the expensive, complex and eco unfriendly print process, mobile communication (i.e. digitized unified communication) does avoid a lot of time consuming and expensive travelling and the electric car (i.e. in the long-run result digitized mobility with no drivers)—as a recent example of the discussion—is said to have zero emissions.

But is this really true and do we consider all dimensions of sustainability? Data storage needs lots of electric power to do its business. And it needs power constantly where a printed paper—i.e. a document once printed—might exists for eons to come and it can be read without complicated machines. And mobile communication might just work fine but it might not bring the same results as a good chat and a glass of wine shared in the evening under foreign or home skies with newly made friends or business partners. Electric power never comes without any carbon footprint. As it is produced at one site, electricity needs lots of transport capacity, landlines etc. and your windmill or solar panel most likely is a product with a vast variety of parts, mostly cheap parts with more or less unknown ecological background. Looking at the electric car, you can be sure that the critical materials used in the batteries and some of the composite parts leave many questions unanswered.

Digitization as such is not the solution to make businesses sustainable.

Sustainable Business

So what is the trick on sustainability meeting the digital world and how can SMEs and Start-Ups position themselves? Some quick considerations:

- Sustainability viewed from the market side is a business imperative, not a maybe. We have to accept the simple fact that customers want to have it. Even if they do not want an originally sustainable product, a huge percentage of customers just feels better when they buy a product that after all offers solid sustainability.
- Viewed from the origin and the definition of the concept, sustainability is multi-dimensional. Simplified these dimensions are ecology, finance and society. Even if the dimensions are equally important, different societies or customers might want to see an emphasis on the one or the other part of the concept—today or over time.
- Sustainability may hold different approaches in different companies as not all dimensions of the concept might equally apply for all products or services and for all creative approaches of any kind of company or appliance.
- Following the understanding of different companies and a broad variety of products and services, sustainability should be approached in that dimension where it has a high impact in the respective company, on the product or service offered and on the relevant stakeholders. This consideration can be extended to the individual spot of the customer experience.
- Sustainability is highly different in its practical approaches in big companies compared to SMEs and Start-Ups. This effect is very important but it is only vaguely understood in sustainability practice.

A short look at sustainability as business imperative:

Businesses continuously derive their license to operate from the societies they supply. The stakeholder approach shows the strategic accessibility and options of this insight. Given the crosslinked relationships between people in their various functions (voters, citizens, employees, customers, entrepreneurs, innovators, etc.), over time (children, parents, ancestors, generations) and throughout region (national, international, global, cross-language, cross-cultural, etc.) sustainability is a business imperative. Businesses which do not follow sustainable paths will finally lack profit, access to resources and people (customers and skilled employees).

A short look at the role of big companies in the field of Sustainability:

In a big company, the mass of the production and its impact (here also the ecological impact) are considered by the market and the stakeholders as it is: it is a big impact and doing bad with a big impact (in ecology) is a bad thing. At a certain point even taking good steps against the bad impact will not solve the problem. The company will be considered being a bad company at this point and this will not change for a very long time. Given the power of eco NGOs or Activist Organisations (AO) (this might be a result of their relatively long time of existence and creativity to reach out to their relevant public) lots of the good sustainability of a big company (e.g. how socially they treat their workforce etc.) will never make it to widespread public knowledge. In the customers perspective an additional effect will happen: some measures taken into sustainability from big companies tend to look weak or faint, with a high probability they will be considered to be greenwashing.

Except sustainability being a general business imperative to all kinds of businesses, compared to Big Companies things can be very different for SMEs. Given the (often) smaller impact dimension of the product created, the work done or the service provided, the dimensions of sustainability postulated from the market (i.e. from customers and along the supply chain) is different. SMEs solve their customers' individual problems, they get more trust from their customers, they are very innovative, they have a higher grade of cooperation and they can respond much faster to (market) demands.

If SMEs and Start-Ups continuously and attentively use innovative methods, new approaches and the cultural change that Megatrends (such as Digitization) bring up, they can improve their market approaches and customer interactions towards the dimensions individual, innovative, trustworthy, cooperative and responsive and they might find a true sustainable business.

It is very important to understand, that SMEs and Start-Ups do not have to change themselves in any core part of their setup or that they do not have to use Trends obsessional in any part of their work. They can selectively use such methods, approaches and change—according to their impact or supply pattern or at such crucial points where the new trend can create positive impact—at the very points and parts where most effect can be provided and they can communicate this selective approach to their relevant customers, too.

If SMEs are part of a supply chain, then they can use the best (i.e. the most sustainable) supplier selectively or on demand, rated individually from the very special strategic sustainability perspective. Good fabricated materials or services can be a solid basis of sustainable production or service creation. Such materials or services can be used wisely (i.e. strategically), maybe on an eco-efficient or socially oriented basis and improve the sustainability of a product and service (in the supply chain).

One additional thing: Many years of experience in the field of sustainability and the structural setup of SMEs show that on the one hand a broader range of sustainability efforts offer more flexibility (towards customer communication) but that on the other hand customers want SMEs to work very precise and focused.

And the final Question: Can a business be sustainable when it is not a champion in the field of ecology? Yes, of course. SMEs and Start-Ups should almost never focus solemnly on ecology, but foster good societal impact on a good financial basis as well. This supports their image of individuality, trustworthiness, innovation, cooperation and flexibility/response speed.

2   Views into a Sustainable Future

In many parts of the (business) world we will see a consequent physical disconnection between workspace, workplace, product and service. As *place* will be disconnected also *time* and *culture* will see the same fate.

How can we envision options for SMEs and Start-Ups in this scenario and include the manifold dimensions of sustainability?

In the way things or concepts are disconnected (decomposed) they can be reconnected (recomposed) again. This can be done in innovative or creative ways. Existing cloud based services can already give us a short glance at the future. And the recent developments in the cloud (i.e. the easy, cheap and stable availability of such services to a broad economic public) illustrate its accessibility not only to big companies but also to SMEs, Start-Ups and creative users. Just like outsourcing was the key concept for (big) companies in former days, companies of any size (and creative users) can now use the same method (outsourcing) with a digital infrastructure (cloud based services or structures).

Employment and Workplace
SMEs and Start-Ups will consequently use cloud based (outsourced) services or solutions as inputs to produce their products or services. Recombining cloud based services or solutions respectively using the digital supply in innovative (new, novel) or creative (unique, artistic) ways will lead to new structures in employment and workplace, too.

SMEs as well as Start-Ups can follow this development and reach out towards a bunch of sustainable solutions.

In the near future any form of permanent office, which is simply too expensive and not flexible enough, will vanish the same way as significant part of permanent employment. If a company can manage this development and foster a corporate culture that has a focus on the needs of the employees and on cooperative structures alongside trust and respect, it can be very successful. Flexible structures open new patterns towards scalability on the company side and on a well-balanced work-life-experience at the employees' or coworkers' basis. This can be a big opportunity to overcome a lack of skilled workers in the demographic transition that lies ahead of our developed societies and it will stabilize economic development, both in (socially) committed enterprises and structurally well-prepared (local or regional) societies.

Cloud Offices and Cloud Work will need a very good digital infrastructure. Sustainability and success for communities—as to offer work and services to their citizens—will highly depend on such infrastructure. At first—when this infrastructure will be built-up (mainly by big companies)—sustainability will see negative effects due to overshooting. On both sides, supply and demand, capacities will most likely be used unwise in that phase. That will cause (short to mid-term) negative eco effects and high prized over capacities or demand in some areas. As other regions will be digitally underdeveloped, parts of our society might not profit from the effects at the same time and rate as others and there can be a further shift towards the big cities for a mid-term time span. SMEs can use their flexibility (adaptiveness) and surely profit from all these effects financially and in many other forms of sustainability, when they compensate time and development gaps and use their resources towards development, access and usage of digital infrastructures and solutions wisely and attentively.

Production of merchandise will see massive change, too. Digitization opens new options in mass customization, order structures and transport paths. Speed factories (Smart Factories) will use this and bring back production from low-income countries to the developed countries that had lost such production in the last decades. But this production will be robot based or use 3D-Printing methods with only a very small human workforce. Smaller companies can use their flexibility and trust positions to profit from this transition process towards digitization based automation: Manufactories can service sophisticated and/or upscale demand (with skilled workers that do not work in automated factories any more) and specialized service providers can solve individual requests. Especially SMEs or Start-Ups will have the opportunity to show very powerful skills e.g. in eco-friendly products, positively motivated employees, etc.

As Digitization will be more and more connected with housing (Smart Home) and the daily life (Internet of Things), products (consumer goods) will be delivered within minutes or in real time after they have been used or ordered. Some consumer goods that people need recurrently will even be provided without previous order just on practical demand as the customer is used to consume them. Sustainability will be challenged from the market, as a 24/7-Society that covers all products and services will have broad social impacts. Flexibility (what, when, where?) reconciled with sustainability (how?) will win the race.

Environment and Energy

Digitization and digital business need a functioning data network. Such kind of businesses additionally need data storage and a lot of (mobile) devices. Network and data availability finally has to be 100% at any time and at any location.

Given a certain amount of overshooting (cp. Employment and Workplace) and capacity effects when services are held in readiness (suppliers, their suppliers, and so on along the supply chain) digital services might not be very eco-friendly at the end. Enterprises who use digital services, networks, storage and devices attentively might profit in two significant ways. First they will save lots of costs, second they will be able to gain competitive advantages in their eco-balance and customer demands of that sort.

Cost pressure will force SMEs and especially Start-Ups to use such services, that offer best value (price, economic benefit, usability, availability) at the same time as best scalability. Customers and investors will reward good setups, even more if associated transport patterns are considered holistically and if they are only used on demand.

Government and Security
Digitization will have big impacts on government structures as well. Besides the effect that citizens will technically know more details on financial, structural or governmental data, lots of digital services will be available. As such services might be part of a fiscal structure (taxes) or users might use them only a few times in their life (construction plan for a house), these services will offer big potential for innovative service providers.

Developments around the Smart Home, the Internet of Things or the Digital Wallet will produce a growing demand on security. It is unclear whether the customers will trust small companies the same way as in the solutions of governmental structures or the solutions of big companies. If SMEs and Start-Ups use transparent approaches (such as open source or open innovation) they can foster the trust component and most likely have a big competitive advantage.

One big benefit might be the barrier free aspect of digitization. Barrier free solutions (not only such solutions that help handicapped people but also such ideas that enable a solid general work life balance) can easily be combined with service components which might open up lots of economic potential.

New Business Models
As Employment and Workspace, Environment and Energy, Government and Security, see change through digitization, entrepreneurs will apply new creative ideas info their business models.

Furthermore, entrepreneurs will enter areas that have only been accessible for governmental or public structures in the past. More and more medical or fiscal services, education and security will be run or supported by private sector professionals.

Cooperation will be key, especially for SMEs and Start-Ups. Any development (as digitization will enable processes to be split up in small parts) will see specialization and a trend towards cooperative (co-)creation of products and services. Open Innovation will speed up such developments. Companies that offer infrastructure (organization of communication and exchange) for such development can profit from these developments. Flexible companies can have lots of opportunities accompanying the change process, e.g. in rural regions, working with the different expectations of various generations or offering solutions for the old economy in their individual transition process.

Successful new business models will most likely be diverse subscription models along the customer relationship. Such models will evolve as customer generations and services change over time or with the product offered. Entrepreneurs will foster their business when they offer more services together with their products. Educational services, such as product trainings, offer big opportunities.

3 Reflections on a Better World

Sustainability is an evolving concept. The same is true for Digitization. As the world is constantly changing, people and societies are adopting and companies do a variety of innovations. After all, Sustainability is the search for a better world, both for people today and people tomorrow. Digitization is a Megatrend that must be accepted as it is (cp. Digitization as Megatrend) but whose evolution, however, can be used or even influenced by creative and innovative economic subjects.

I have the express vision that business done with embedded, resilient and cooperative structures will be very successful over time.

Embedded Sustainability (Sustainable Digitization)
At the beginning of the sustainability voyage, sustainability was an add-on to "real business". Eco efficiency, good relations to your employees and a more or less functioning public environment have been added in those times to construct a greener product, better employee or public relations and a premium priced product.

With customers using social media to exchange their experience and opinion the starting situation has changed. Flat information hierarchies (driven by Digitization)—even on formerly top secret company data (invested capital, machines in use, average wages, etc.)—and the understanding that customers made sustainability a business imperative, sustainability cannot be a simple add-on any more.

If—in the daily company practice—Sustainability is not an add-on any more (that can be disconnected from the core company purpose) but it is considered to be an integral part of the core business, two effects will start: First Sustainability will be a competitive advantage (ether on the supply side, e.g. trough reduced cost structures, or on the demand side, e.g. fulfilment of the customers' implicit demand for ecologically or socially good products). Second the company will profit from its Sustainability being part of the value creation.

Usually a success factor will be developed and extended over time. Sustainability can be understood as a role model for a concept which—if done truely embedded (Laszlo/Zhexembayeva 2011) in a systems' core—can result in a success spiral. Good Sustainability will lead to more good sustainability. I understand that the same is true for Digitization or—even better—for Sustainable Digitization.

Resilience (Resilient Digitization)
Entrepreneurs who follow sustainability pathways most likely will find resilience for their businesses as well.

Resilience can be understood as a (corporate) setup where a system (company) is established in such a way that it can cope with a shock (or dramatic change) scenario. Resilient systems are prepared to suffer a shock, maybe stand it or recover very fast respectively emerge with renewed strength after tripping.

Sustainability supports resilience as it creates strong relationships and trust (e.g. with your employees, suppliers or the relevant public or other stakeholders). In difficult economic times these relationships are more reliable, stable and durable.

Cooperative Innovation (Cooperative Digitization)

Innovation is disruptive in its core. Following the Schumpeterian logic of Creative Destruction, something old will be replaced and finally destroyed by something new. As Megatrends, such as Digitization, shape (change) our world in an unprecedented speed and outreach, also the disruptive effects speed up. The way to new equilibria is (felt to be) disruptive for those who (short term) suffer from that change (e.g. as they lose their jobs) and for such systems and setups that are destroyed by the new methods and approaches.

Using Digitization, companies and innovators of any size alike foster their strategic or financial goals—neither caring for existing equilibria nor estimating the costs on the side of those being disrupted. Without suitable attentiveness business schools, innovation hubs and even political protagonists propagate disruptive innovation as key to a bright future (cp. cooperative-innovation.com).

Cooperative Innovation (Wenzel 2014) offers a paradigm shift for such scenarios as it includes all actors and combines attentively the best parts of existing and new ideas into a new solution that does not produce losers. Where disruptive innovators enter the arena in a fighting mood and by force disrupt the existing work, trade or lifestyle equilibrium, cooperative innovators show their innovation to the relevant public and try to include existing solutions and the new approach into one superior solution. Thus no fight is necessary, war costs are minimized or eliminated and the disruptive effect of the new idea might come to its full potential faster, cheaper and without leaving others behind respectively without risking the loss of a functioning and so far resilient system (cp. cooperative-innovation.com).

References

Laszlo C, Zhexembayeva N (2011) Embedded sustainability: the next big competitive advantage, 1st Ed. Stanford University Press, Stanford

Naisbitt J (1982) Megatrends: ten new directions transforming our lives, 1st Ed. Warner Books, New York

Smith A (1776) An inquiry into the nature and causes of the wealth of nations, 1st Ed. Strahan Cadell, London

Wenzel F (2014) Cooperative innovation: paradigm shift for disruptive innovators towards sustainable digital business. https://www.cooperative-innovation.com. Accessed 26 Nov 2016

Sustainable Cooperate Information Portals: Digital Knowledge Communities for SME

Martin Kreeb and Hans-Dietrich Haasis

1 Construction of the Ecoradar Knowledge-Community

A large variety of research has been published in the field of Sustainable management during the last 20 years (Baumgartner and Rauter 2017). The problem was the conversion of this knowledge (Loebbecke and Myers 2016) into enterprise practice Development-Target of the ecoradar-portal is it to reduce the information costs of those SME enterprises, which are interested in Sustainable management. Especially in the subjects of energy management and climate protection instruments. In order to achieve these targets, a strategic Community concept of the third generation has been developed in order to build a knowledge-community (Deshpande et al. 2017) in the SME sector.

The main emphasis of the ecoradar-community is on the knowledge field and the service and project-areas. The community will start as a project-community. In the beginning, ecoradar, as a classical research project, is measuring the success by certain criteria focusing on timeframe and milestones (Vlas et al. 2017), evolutionary Software Requirements Factors and their Effect on Open Source Project Attractiveness. In the Proceedings of the 50th Hawaii International Conference on System Sciences, an additional feature is the use of a virtual project team (scientists, consultants, entrepreneurs). A virtual cooperation has been realized by establishing a specific editorship- and tele-cooperations system. This project-communities represent the preliminary stage on the way to a knowledge-

M. Kreeb (✉)
Fresenius Business School, Munich, Germany
e-mail: kreeb@hs-fresenius.de

H.-D. Haasis
Bremen University, Bremen, Germany
e-mail: haasis@uni-bremen.de

© Springer International Publishing AG 2017
T. Osburg, C. Lohrmann (eds.), *Sustainability in a Digital World*, CSR,
Sustainability, Ethics & Governance, DOI 10.1007/978-3-319-54603-2_12

community. Ecoradar will be a knowledge network stretched beyond the limits of individual organizations and enterprises (Krämer and Kalka 2017).

Wenger and Snyder 2000 describe the knowledge-community as a "flexible organizational unit, beyond official organizational resp. informal units. The community is animated by the common interest of the members in the field of knowledge. The participation is voluntary. The motivation to participate is a positive cost/benefit relation." (Wenger and Snyder 2000).

The collective benefit is categorized by Rheingold (Rheingold 1994) using the following three dimensions:

- Social use, identification by a common goal
- knowledge capital, use of knowledge from various sources
- community feeling, system of real contacts and experience backgrounds

The ecoradar-community understands itself as community of interests, with the following features defined by Hagel and Armstrong (1997: 23)

- focus and emphasis on a specific interest
- the ability to integrate contents and communication
- the use of information, supplied by the members
- the access to competing providers

The major task of the community-developers is the professional relations management between the individual community-members. The goal of the ecoradar-relation management is to integrate stakeholders like NGO, companies und communal administration in the community process. This means that anonymous coworker will be transformed into active community-members. The socio-economic-group-dynamic pocesses together with technological-organizational processes have absolute priority.

2 Knowledge Management in the Ecoradar-Community

For the joint-project an expert set of 21 different research institutions could be won. The expert set has the function to edit the relevant knowledge of the "community-environment" so that enterprises can transfer this expert knowledge to the Sustainable-oriented management. The knowledge management model of ecoradar supports the creation of knowledge within the enterprise on the basis of the external source of knowledge in the sense of the ontological knowledge spiral. The expert knowledge helps to support the acquisition of external knowledge and the development of own knowledge. The actual knowledge distribution is supported both over a especially designed telecooperations-system as well as over the portal (Frömmgen et al. 2016; Böhmann and Krcmar 1999). That telecooperations-modell as well as the portal is regularly updated by the experts and is supporting the knowledge preservation (Bannister and Grönlund 2017) in the organization. In the later course of the project it has to be assessed by the experts whether a ontology-based knowledge evaluation can be realized (Huang et al. 2017). The evaluation research in co-operation with enterprise practice and with the help of empirical methods has to ensure that the quality criteria that are pursued by ecoradar such as Sustainable discharge, target group orientation and in particular practice fitness (glossary word: SME proximity) are actually respected and realized. The evaluation of enterprise practice will be performed by the practice-community.

The development team of ecoradar confirms the experience of Davenport and Prusak (1998: 32), that knowledge can exclusively be created in the brains of the knowledge carriers. The knowledge carriers of ecoradar are scientific experts and entrepreneurs, who cooperate within the community-process. The primary focus is on the externalization of the expert's knowledge. The know-how is transferred in an external information system (Knowledge Warehouse, CMS). Externalization of knowledge (Zhou et al. 2017) is especially suitable for standardizable knowledge (standards, laws, etc.). The recent experience of the ecoradar research project has shown that direct communication in a Knowledge Network is the best way to convey the expert's knowledge and experience (see below Table 1).

Table 1 Knowledge warehouse versus knowledge network (own illustration)

Criteria	Knowledge Warehouse	Knowledge network
Philosophy	Externalisation of knowledge	Direct communication, Reference to human experts
Range of application	Structured problem areasgiven goalknown relevance of informationConsequences of the decision foreseeablere-usable solutions	unstructured problem areasnot given goalunknown interdependencies Consequences of the decision unforeseeable limited reusability of solutions
Artificial intelligence	High (CMS)	low
Knowledge-requirements	Rules and methods	Not exactly specifiable
Moment of knowledge division	at the beginning of the knowledge process	On demand
Method to display knowledge	structured knowledge	Reference to knowledge carriers as well as presentations of experts's assessment
Knowledge transfer	Knowledge conveyed by knowledge carrier (experts)	Bilateral negotiating of the modalities for the sharing of knowledge
Role of IT	Storage and processing of knowledge	support of the information process and communication process
Access to knowledge	Information Retrieval & Data Mining (Sathiyamoorthi 2017)	creating of contact and communication with knowledge carrier

3   The Ecoradar Practise-Community

Representatives of the joint project's target group, enterprises in Germany, have already given it broad approval in its start-up phase. Some 40 enterprises employing an estimated one million members of staff have made the decision to support production and development of the prototype. The development of so-called 'eco-radar' screens is to be carried out in 18 workshops, hand in hand with business representatives and numerous experts. The organization of the high-calibre working groups has been taken on by Europe's largest business-led Sustainable initiative, the German Sustainable Management Association (BAUM e.V.), Hamburg. In addition, in the summer of 2001 a representative written survey was conducted in around 9000 enterprises-ECORADAR is a prototype early detection system which will enable German enterprises to identify technical, political and economic risks—but also market opportunities—of an Sustainable nature much earlier than their competitors, and to assess them more competently.

4   Content-Model

The ECORADAR system portal consists of eight ECORADAR screens which users can view as an ensemble—or individually if preferred—to scan a company profile (Company Radar—'micro-level') or the wider economic setting (Macro Radar—'macro-level'). The Company Radar is a system component that can be accessed from any ECORADAR screen, enabling users to systematically record and evaluate their company Sustainable Data, their company Sustainable Policy and their company Sustainable Goals. The Macro Radar, a similar system component that can be accessed from any ECORADAR screen, enables users to record and evaluate the 'macro-level' on the basis of the latest research—for instance global, national and regional Sustainable Data and Sustainable Goals as well a Content Management.

Within the project ECORADAR there will be created a portal that supplies Sustainable services. First, it is essential to embed information, references and checklists that have been already part of the ECORADAR-FRAMEWORK and former designs. In addition to these functions, the final version should be able to support all users interested in the Sustainable field by providing a virtual community. It should also identify possibilities for cooperation between all participants. Finally, it should enable the integration of Sustainable Management in business processes.

The first step is the creation of a user-friendly page layout. The essentials are a clear graphical structure, simple usability and the direct access to the services that are available with short download times.

5 Content Structure

ECORADAR is the result of a wealth of research which has mounted up over at least two decades. There are copious research findings under all eight of the sub-headings, along with applications that have been tested in practice, in some cases. Some parts of the ECORADAR system rely heavily on the latest Sustainable performance standards. The ECORADAR sequence of 'Sustainable Data—Sustainable Policy—Sustainable Goals—Sustainable Organization—Sustainable Knowledge' largely follows the thought processes of the European Union Eco-Management and Audit Scheme (EMAS) and ISO 14001. The integration of the ECORADAR screens 'Sustainable Costs', 'Sustainable Market' and 'Sustainable Technology' in the overall system is largely attributable to experience reported by companies. In business practice apparently there is plainly a recurring need for this kind of information.

5.1 Guidelines for Action

ECORADAR is a prototype early detection system which will enable German enterprises to identify technical, political and economic risks—but also market opportunities—of an Sustainable nature much earlier than their competitors, and to assess them more competently. The ECORADAR system portal consists of eight ECORADAR screens which users can view as an ensemble—or individually if preferred—to scan a company profile (Company Radar—'micro-level') or the wider economic setting (Macro Radar—'macro-level').

Company Radar
   The Company Radar is a system component that can be accessed from any ECORADAR screen, enabling users to systematically record and evaluate their company Sustainable Data, their company Sustainable Policy and their company Sustainable Goals.

Macro Radar
   The Macro Radar, a similar system component that can be accessed from any ECORADAR screen, enables users to record and evaluate the 'macro-level' on the basis of the latest research—for instance global, national and regional Sustainable Data and Sustainable Goals.

5.2 Four-Point Menu for the Company Radar

ECORADAR will use the Internet to provide structured communication of the latest expertise on sustainable management in a way that assists decision making and is comprehensible and relevant to enterprises. A four-point menu—which once again is integrated into all ECORADAR screens—will ease this task for companies.

1. 'Getting Started'
   The 'Getting Started' menu shows companies the fundamental points they should take into account.
2. 'Stumbling Blocks'
   The 'Stumbling Blocks' menu shows how common mistakes can be avoided.
3. 'Checklists'
   The 'Checklists' contain guidelines for action which can be used interactively.
4. 'Benchmarks'
   The 'Benchmarks' allow for comparison with other enterprises by 'looking over their shoulder'.

6 Portal Structure

ECORADAR is the result of a wealth of research which has mounted up over at least two decades. There are copious research findings under all eight of the sub-headings, along with applications that have been tested in practice, in some cases. Some parts of the ECORADAR system rely heavily on the latest Sustainable performance standards. The ECORADAR sequence of 'Sustainable Data—Sustainable Policy—Sustainable Goals—Sustainable Organization—Sustainable Knowledge' largely follows the thought processes of the European Union Eco-Management and Audit Scheme (EMAS) and ISO 14001. The integration of the ECORADAR screens 'Sustainable Costs', 'Sustainable Market' and 'Sustainable Technology' in the overall system is largely attributable to experience reported by companies. In business practice apparently there is plainly a recurring need for this kind of information.

6.1 Sustainable Data

Sustainable data are generally held to be the 'oxygen' of Sustainable policy. The regional, national and global Sustainable data provide a key basis on which companies can take action. Wherever the Sustainable situation is monitored and observed, wherever citizens are surveyed on their subjective experience of Sustainable problems, this can provide the impetus for action in Sustainable policy. Elementary company Sustainable data, for example, might be figures relating to energy, water, wastewater, waste, emissions and hazardous substances. Carbon dioxide emissions would be one example of key global Sustainable data.

6.2   Sustainable Policy

Approaches for Action Towards Sustainable Management
The future Sustainable standards imposed on enterprises are moulded partly by their own Sustainable policies but especially by external government and party programmes. For example, national environment policy approaches for action form an important basis for the future use of 'command-and-control' instruments. In Germany, for instance, the ideas of the coalition parties, the opposition and the separate parties at national, federal state and municipal level are not the only matters of importance. A considerable influence is exerted on future Sustainable policy by the policy-making bodies of the European Union and numerous other international organizations.

6.3   Sustainable Goals

Principles for Action Towards Sustainable Management
While Sustainable data represent a significant basis on which to take Sustainable policy action, Sustainable goals provide principles for action which, for their part, form the basis for the future application of environment policy instruments. Society should come together and use environment quality objectives to define core elements of environment policy action, working towards sustainable management in years to come. A company's own Sustainable targets, in contrast, are an element of the internal early detection system. Basically these should be geared to continuous improvement of Sustainable performance.

6.4   Sustainable Organization

An effective Sustainable early detection system can only be incorporated successfully within the enterprise once an efficient organization is in place for the structure and processes of Sustainable performance. Because then, and only then, is it possible to perform the target-performance comparisons which are necessary for early detection. For early detection, another important factor is to work closely with the public Sustainable authorities and associations: Sustainable authorities are the pivotal interface between the letter of the law and its enforcement. Enterprises that maintain good contacts with Sustainable authorities have swift access to information on new requirements under Sustainable law. Associations are viewed as powerful Sustainable policy actors and can pass on to their corporate members targeted advance information on Sustainable performance, picked up during the course of their lobbying.

6.5   Sustainable Knowledge Management

Sustainable know-how, both inside and outside a company, is a central element of Sustainable early detection. A cornerstone for knowledge transfer in the Sustainable sphere is formed by institutions such as the German Federal Sustainable Agency, the Federal Agency for Nature Conservation, the Federal German Foundation for the Environment, and the International Transfer Centre for Sustainable Technology. Likewise the media, as environment policy opinion-formers, play an important part in early detection.

6.6   Sustainable Costs

Monitoring and assessment of Sustainable costs in the widest sense (calculation of a company's pollution control costs, anticipation of external costs and the costs of neglecting Sustainable aspects, identification of potential cost reductions) is a permanent task within early detection. In particular, deducting—at least mentally—the costs of Sustainable degradation (today's external costs—tomorrow's operating costs) is a strategic element of eco-controlling.

6.7   Sustainable Market

Sustainable protection has developed into a significant economic factor over the past 30 years. In the year 1997 alone, German private and public sector spending on Sustainable protection was around DM 65,000 million. Studies predict that the market for Sustainable technology and Sustainably friendly products will continue to grow internationally in the coming years. Admittedly Germany still has a high market share in this area. However, other industrial nations—notably the USA, Canada and Great Britain—have developed strategies for gaining targeted access to new markets and supporting exports of Sustainable technology by their suppliers.

6.8   Sustainable Technology

Technical indicators play an important part in the early detection process. In particular, specialist trade fairs and exhibitions not only forge new contacts and stabilize business relationships but also provide advance information on technical innovations. Delphi surveys are increasingly conducted as part of this technology preview process, and these can serve to guide future strategic orientation.

7   Information Technology

Information technology (IT) research should contribute to ensuring that ECORADAR actually fulfils the quality criteria it has set itself, namely coherence and effectiveness, capacity for integration, clarity and, in particular, user-friendliness. The ECORADAR system must measure up to the latest developments in IT so that it can do full justice to its future-oriented role. Intelligent solutions must be developed for three fields in particular:

7.1   ECORADAR as a Workable Tool

The concern here is to create interactive, creative opportunities for the user (examples: automatic generation of indexes on the basis of a personal database; form-filling assistance; checklist programmes). The success of the ECORADAR system may well critically depend on the level of convenience built into the system architecture.

7.2   Integrating ECORADAR into Existing Business Processes

The better Sustainable performance is integrated into typical business procedures, the greater the prospects of success for sustainable management.

7.3   Mounting ECORADAR Technology on the Internet

The core parts of the ECORADAR system should be placed on the Internet as soon as possible (no later than 1 year into the project) and continually updated so that the feedback coming from users can be integrated reasonably quickly into the current research and development process. ECORADAR forms an ideal foundation for an Internet portal for 'sustainable management' and can be seen as the seed from which such a portal may grow.

8   Internet Strategy

The concept of a web portal has proven to be useful to handle the overwhelming data available on the internet. A portal can structure the information and is able to display the content in a user-friendly layout. This is the basis for an effective research by the business community. A portal is a universal and comfortable system to access applications, content and services that are focused on a specific topic.

Portals can be labeled as web-based: multimedia-style and accessible via standard internet-browsers

- task-oriented: adaptable regarding the tasks of users or customers
- categorized: content and services structured by categories
- personalized: individually designed to achieve 1:1 relationships with users/ customers.

9   Internal and External Aspects of the Portal

The original concept of portals (i.e. Yahoo) was focused on the private, individual internet user. The main difference between a portal and a search engine (i.e. Google) is, that experts prove the content—not mathematical algorithm like the Google information world.

The idea of the portal is now increasingly focusing on individual companies. This is called an "Enterprise Information Portal" (EIP) (Kumar and García 2017). An EIP is focused both on internal users (employees and management) and external parties (customers, suppliers and other stakeholders of the company).

The internal focus of the portal has increasingly been on knowledge-management and the supply of software applications (i.e. inventory management, PPS, sales).

The external focus has in addition also functions for transactions like e-commerce, e-procurement, e-logistics and supply-chain-management). The internal interface is sometimes referred to as "Workplace", while the external side is called "Marketplace" (see SAP AG, mySAP.com). The themes of a portal, like applications, content and services can be designed to suit the needs of a specific geographical region or enterprise and the themes can also be selected to cover the requirements of a specific task or problem. It is also possible to mix a focus of a specific subject and a specific enterprise.

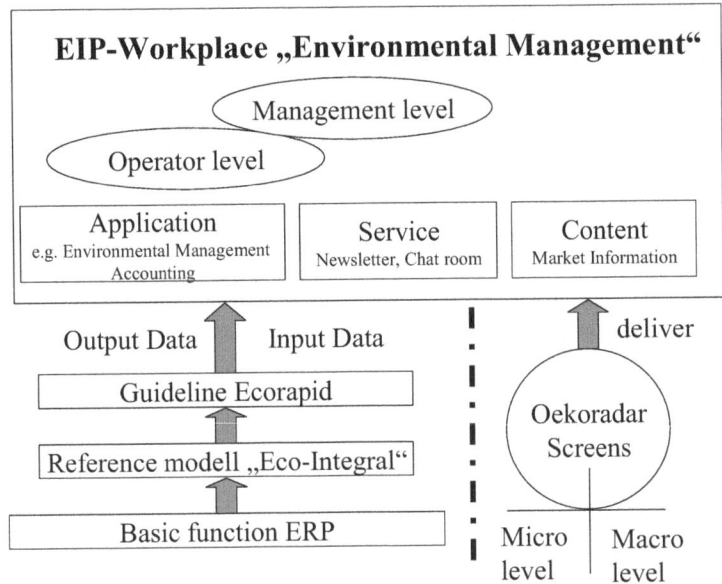

Fig. 1 EIP-workplace "sustainable management" (own illustration)

10 Workplace and Marketplace-Functions

The basic idea of ECORADAR (Kreeb et al. 2009) is the combination of "Enterprise Radar" and "Surrounding Field Radar". This is the ideal basis to create a theme-related portal with a public/external side ("Marketplace") to supply content and services for all companies and individuals that are interested in "Sustainable Management" and an internal side ("Workplace") to supply the enterprise with functions for "Sustainable-Management" with both strategic and operational tasks.

The following illustration (Fig. 1) will show this internet-based dual approach:

The key innovation is the consistent use of all available "internet technologies". The great idea of sustainable management will substantially benefit from this transition towards "Internet-Economy".

11 Creation of an Internet-Platform for Sustainable Management: A "Workplace"-Architecture

One of the important trends in Enterprise-Data-Processing is the introduction of the so-called "Enterprise Information Portals" (EIP). As described above, every individual employee (from a simple operator to top managers) is offered customized information, applications and services via a common, open-platform internet-browser. Beside these operational functions there are also now increasingly offered various functions in knowledge-management. The final goal of the development of a portal based on ECORADAR is the concept and consecutive design of a workplace (following the concept of an EIP), that offers all information, applications and services necessary for the tasks in Sustainable Management.

In addition to ECORADAR, the project "ECO-Rapid" is also an important basis for this development. Both projects are cooperating. There is an active exchange of results and planning. The following illustration is roughly showing the workplace-architecture on the basis of "ECORADAR" and "ECO-Rapid": Creation of a web-based, open, dynamic access to all relevant Resources of Sustainable Management for Enterprises and Individuals. The bundling of all relevant resources concerning Sustainability on one webpage is the primary goal of this public portal. The task of this portal is to cover all needs of enterprises and individuals for information about the topic of Sustainability. There is a substantial demand for that kind of bundled information in Germany.

The following topics are possible and some are already integrated in the presented framework of the prototype ECORADAR:

- current and historic Sustainable data
- knowledge base for Sustainable Management and
- Sustainable Technology
- Sustainable Laws, intelligent checklists for individual use
- Ecologic Market (Purchase of ecological products for enterprises and private households)
- Ecologic investments
- List of ecologic business consultants
- Specific literature

The portal is offering three main functions:

1. Passive, regularly updated information for research
2. Interactive communication between users, assuming that there is a demand for exchange of specific subjects via chat-rooms, interactive message-boards, exchange of knowledge and experiences
3. Transactions, products and services. The portal can be upgraded for electronic procurement of Sustainably friendly products and services.

The module for supply of information can already be almost completely covered by ECORADAR. It would be required to create a Content-Management-System that provides a regular flow of information at reasonable costs. The module for transactions could be started with partner-companies and then be gradually expanded.

12   The Editorial System and Telecooperation-System

The design of the portal is also requiring the development of a technological infrastructure. The community that is providing the Sustainable information needs a system for editing and telecooperation. The careful design of a sustainable project has to ensure the possibility for current and easy upgrades. The approach of iterative prototyping and learning by-doing will provide constant input by users that can be integrated in the development process. The valuable data will also contribute to the development of a business model for the portal ECORADAR. Data about addresses, for advertising, newsletters and commissions is essential.

References

Bannister F, Grönlund Å (2017) Information technology and government research: a brief history. In: Proceedings of the 50th Hawaii international conference on system sciences, Hawaii

Baumgartner RJ, Rauter R (2017) Strategic perspectives of corporate sustainability management to develop a sustainable organization. J Clean Prod 140: 81–92

Böhmann T, Krcmar H (1999) Werkzeuge für das Wissensmanagement. In: Antoni CH, Sommerlatte T (Hrsg.) Spezialreport Wissensmanagement—Wie deutsche Firmen ihr Wissen profitabel machen. Symposion Publ, Düsseldorf

Davenport TH, Prusak L (1998) Working Knowledge. Harvard Business Press, Boston

Deshpande DS, Kulkarni PR, Metkewar PS (2017) Need of the research community: Open source solution for research knowledge management. In: Open source solutions for knowledge management and technological ecosystems. IGI Global, Hershey, p 146–174

Frömmgen A, Heuschkel J, Jahnke P, Cuozzo F, Schweizer I, Eugster P, Buchmann A (2016) Crowdsourcing measurements of mobile network performance and mobility during a large scale event. In: International conference on passive and active network measurement. Springer International Publishing, Heidelberg, p 70–82

Hagel J, Armstrong AG (1997) Net Gain. Profit im Netz: Märkte erobern mit virtuellen Communities. Gabler, Wiesbaden

Huang JS, Kozaki K, Kumazawa T (2017) Knowledge structuring for sustainable development and the Hozo tool. In: Open source solutions for knowledge management and technological ecosystems. IGI Global, Hershey, p 195–221

Krämer A, Kalka R (2017) How digital disruption changes pricing strategies and price models. In: Phantom Ex Machina. Springer, Heidelberg, p 87–103

Kreeb M, Dold G, Haasis H-D (2009) ECORadar-Shakti—an interactive knowledge-base contributing to the[nbsp] greening of an Indian megacity. In: Hallin A, Karrbom-Gustavsson T (eds) Organizational communication and sustainable development—ICTs for mobility. IGI Global, Hershey

Kumar V, García DM (2017) Introduction to enterprise portals. In: Beginning Oracle WebCenter Portal 12c. Apress, New York, p 1–5

Loebbecke C, Myers MD (2016) Deploying internal knowledge portals: three major challenges. Information & Management, Amsterdam

Rheingold H (1994) Virtuelle Gemeinschaft. Soziale Beziehungen im Zeitalters des Computers. Addison-Wesley, Bonn

Sathiyamoorthi V (2017) Data mining and data warehousing: introduction to data mining and data warehousing. In: Web data mining and the development of knowledge-based decision support systems. IGI Global, Hershey, p 312–337

Vlas R, Robinson W, Vlas C (2017) Evolutionary software requirements factors and their effect on open source project attractiveness. In: Proceedings of the 50th Hawaii international conference on system sciences, Hawaii

Wenger E, Snyder WM (2000) Communities of practise. The organizational frontier. Harv Bus Rev 78(0–1):139–145

Zhou W, Yan W, Zhang X (2017) Collaboration for success in crowdsourced innovation projects: knowledge creation, team diversity, and tacit coordination. In: Proceedings of the 50th Hawaii international conference on system sciences, Haweii

Digital Fuel for the Mobility Revolution: The Opportunities and Risks of Applying Digital Technologies to the Mobility Sector

Stephan Rammler

1 Introduction

This article describes the relationship between digitalization and sustainability, taking mobility as an example. We distinguish five innovation paths with regard to the digitalization of mobility. Together, these innovations are expected to deliver massive gains in efficiency and traffic safety. But as well as opportunities, there are also risks. Following this introduction and a definition of terms (Sect. 2), we outline the innovation paths along which digitalization can proceed in the area of transportation (Sect. 3) and the risks that are involved (Sect. 4). Finally, we summarize our findings and draw some conclusions (Sect. 5).

2 Definitions

In a technical sense, "digitalization" is the process of using computer-aided information and communication technologies for calculating, supporting, controlling and connecting processes, procedures and product systems. More generally, digitalization can be described as a phenomenon relating to a new era of civilization: the increasing pervasion of digital systems in all areas of knowledge and life. The "digital society" does not replace phenomena from previous eras, such as industrialization or the service culture, but rather puts them in the broader context of a digital culture. Industrial production processes continue to exist, as does a service economy, spurred on and enhanced by digital media.

S. Rammler (✉)
Hochschule für Bildende Künste, Braunschweig, Germany
e-mail: st.rammler@hbk-bs.de

© Springer International Publishing AG 2017
T. Osburg, C. Lohrmann (eds.), *Sustainability in a Digital World*, CSR, Sustainability, Ethics & Governance, DOI 10.1007/978-3-319-54603-2_13

To use the language of "innovation economics", the evolution from an industrial society to a service society, and thence to a digital society, is as a movement from innovating in the way we produce things to innovating in the way we use things, and thence to innovating whole systems. For the first time in history, today's digital technologies enable us to overcome the—mainly material—limitations of large-scale industrial infrastructure systems and to carry out far-reaching cross-segmental innovations. A good example of this is the coming together of the energy, mobility and communication segments in a "smart grid", resulting in an enormous need for user and other interfaces, such as smarter electric vehicle charging stations.

Digital technologies may also be seen as *general purpose technologies*—technologies that can be universally applied and whose areas of use include the whole spectrum of individual and societal needs and activities.

3   Innovation Paths

In this section we examine current developments and key future expectations in different innovation paths or "clusters" within the mobility sector.

3.1   *Intra-Modal Interconnectivity: Connected Driving*

For the past 10 years, the focus has been mainly on in-vehicle interconnectivity and digitalization. However, the "next big thing" in the automotive industry is widely thought to be *connected driving*, or connecting the vehicle to the act of driving—which is also a precondition for automated driving. Essentially, this means integrating vehicles into the Internet of Things and linking them up to an intelligent traffic infrastructure ("Car2X") and to other vehicles ("Car2Car"). Sharing real-time position and condition data about the driver, the vehicle and the driving situation serves to optimize the act of driving and the traffic flow. This, in turn, enhances efficiency (making optimal use of the infrastructure, avoiding traffic jams, equalizing traffic flows, finding parking spaces) and security. Related aspects include optimizing navigation processes and on-board infotainment.

The intra-modal interconnectivity of the automobility system is thus closely linked to the older concept of "traffic telematics"—intelligent urban traffic guidance systems, parking-space management systems, and so on. While traffic telematics depended on the evaluation of indirect data (sensors monitoring traffic density, people reporting traffic jams, information from road users), in the future, permanent real-time access to the total picture of the interconnected traffic flow will have enormous potential for optimizing stationary and moving traffic, as well as traffic safety.

Today, high-potential intra-modal interconnectivity is finding its way not only into the system of automobility but also into other traffic systems, such as rail transportation. Here, for example, the intervals between trains can be reduced to make better use of lines.

3.2 Intermodal Interconnectivity: Connected Mobility

"Seamless mobility"—integrated mobility, as frictionless and seamless as possible, also called "inter-modularity" or "multi-modularity"—has long been considered a visionary concept of systemic traffic optimization. Today, with the possibility of connecting a wide range of devices digitally, it is becoming a real option. No longer will we rely on one single means of transportation, such as the automobile in Europe and North America. Increasingly, we will intelligently combine various modes of transportation, each with its own advantages and disadvantages.

Digital media, with their broad spectrum of possible applications, play the role of "technical integrator", providing information, matching systems and creating transition points and integrated accounting systems. The ubiquitous smartphone increasingly acts as a "killer application" for new mobility services. In conjunction with ever cheaper data flat rates, smartphones are helping significantly reduce transaction costs and times for using mobility services. User interfaces are becoming more and more intuitive and user-specific, lending a playful, ostentatious character to consumption.

From a supplier's perspective, digitalization makes it possible to interconnect various modes of transportation by providing integrated advance information, planning, booking, access, on-trip information and billing. For the first time in history, digital technologies allow users to plan, carry out and revise multistage journeys in real time. At present the smartphone is the principal terminal, but soon Google Glass, some sort of acoustic device, or a combination of the two could be used. Projects such as Qixxit, Moovel, or the Smile mobility platform of the Vienna public transit network Wiener Linien are good examples of integration concepts. Their weak point is that each of them is designed and marketed by a single supplier. The fundamental question is thus also how to cooperate in a competitive traffic market. Who owns the customer and, more importantly, who owns the customer's data?

3.3 Navigation

There can be no mobility without navigation. The importance of navigation for developing mobility cannot be overrated. While everyone is talking about drive technology and infrastructure, navigation is an area that tends to be neglected. But if you do not know where you are, or what the path from A to B looks like, no movement can take place.

Navigation increasingly plays a central role in modern, complex societies. The simpler the processes of navigation become, the more mobility can be optimized, including in terms of its sustainability. The progressive digitalization of navigation enables—indeed requires—us to orient ourselves in three types of worlds: the real word, the world of digital data, and between them in a mixed world of digital and geographical landmarks, moving objects and people increasingly endowed with a virtual layer of significance. Technological progress has made the virtual world of the global Web, data clouds and digital parallel worlds so complex that we may soon depend on individual "route scouts" and research assistants simply in order to orient ourselves.

3.4 Infotainment

Customers increasingly take powerful infotainment environments for granted. As with navigation, smartphones and other mobile terminals drive the integration of the different functions, and equipment is no longer permanently built into vehicles. Especially in large urban areas in Asia, infotainment systems combined with standard communication technologies play an increasingly important role for customers, improving the quality of time spent in traffic jams and allowing drivers to use the time stuck in their vehicles for work, information, communication or entertainment purposes.

3.5 Telecommuting, Telepresence and Virtualization

In this innovation path, the focus is on the possibility of reducing the need to travel. How can face-to-face interactions be replaced by telepresence, such as video or teleconferences, thereby minimizing transportation time and costs? Conceivable areas of application include telemedicine, teleworking, telelearning (e.g., massive open online courses, or MOOCs), teleshopping, telebanking and possibly *cyber tourism*, or virtual travel for recreational purposes.

"Virtual mobility" refers to forms of mobility where the person does not physically move from one place to another, but uses information technology to travel in virtual reality. This process of "virtualization" has two types of effects:

– *Inductive effects of virtualization*

An increasing proportion of people's individual shopping activities are shifting online. Examples include banking, travel bookings, and ordering products and services. Where established patterns of movement, such as shopping for food and clothes, are replaced by deliveries of ordered goods that more than compensate for the original routes in terms of quantity and quality, these new forms of business have the effect of actually creating more traffic. The dramatic increase in CEP (courier, express, parcel) services in urban areas, for instance, can largely be attributed to changes in ordering behavior, i.e., e-commerce.

The inductive effects of virtualization can also occur on a more long-term, indirect or hidden level. Examples include cheap global communication opportunities facilitated by the Internet. These help develop and maintain a global network of interaction, ultimately resulting in a trans-nationalization of personal biographies. By using virtual media to project their lives into a global space of activity, people are helping open, change and stabilize this space for others, especially for coming generations. As long as the general economic and technological conditions in the global traffic system do not deteriorate, this leads to the development of a global network of traffic-inducing friendship- and relationship-based connections, growing denser with each generation, already no longer limited to global elites but also including the global middle class as a matter of course.

– *Substitutive effects of virtualization*

The opposite of this is to actually avoid traffic. The best example is international business traffic. Videoconferencing technologies, Skype and other means of communication increasingly enable people to replace real face-to-face interaction with telepresent real-time communication. While this option is not suitable for all forms of exchange and negotiation, current developments illustrate the many and varied ways in which online conference software can be used—by engineers to carry out joint construction and product-design processes, by researchers to stage scientific conferences, by globally operating companies to perform regular management workshops, and so on. When, in 2010, the Icelandic volcano Eyjafjallajökull paralyzed air traffic for several days, many companies switched to videoconferencing technology (which, interestingly, the companies had already installed). Prompted by the enormous savings, many of them have since continued to use this technology, permanently reducing their travel expenses.

Videoconferences in particular have significantly greater "media richness", that is to say they are much better suited to creating and maintaining social closeness, than other forms of telecommunication. This makes them an ideal substitute for face-to-face communication. As more linguistic and visual information—facial expressions, gestures, behaviors, the general appearance of the dialog partners, etc.—is transmitted than with other media, a greater social closeness develops compared to related applications such as audio conferences, e-mail or chat.

The main drivers of today's renaissance of videoconferencing systems, apart from the enormous technological advance, are short-term crises (e.g., a fear of flying, flights cancelled after terror attacks, epidemics, volcanic outbursts) and long-term crisis situations (e.g., economic recession). But by proving their value as fall-back options and cost-reduction measures, these systems have made the leap from widely discussed future technology to daily business routine. Video technology has proven especially suitable for internal meetings, which account for up to 40 percent of the travel budget of multinational corporations.

Virtualization is not limited to passenger traffic, either; it also has great potential for goods traffic. Following the marginalization of traditional products in the music industry (i.e., records and CDs), the movie business and the book and newspaper industries in particular are now facing similar developments. Interestingly, strong opposition still comes from the otherwise progressive scientific community, where publications in e-books are not seen as equivalent to print publications.

3.6   Conclusion: A Digital Wave Is Engulfing Mobility

The state of the public debate, as well as the investment volumes and current strategies seen in the automotive and IT industries, show a clear focus on the intra-modal optimization of the automobile system, that is, on increasing vehicle interconnectivity through assisted and automated driving. There is much evidence to show that the long-standing success of established automotive manufacturers with their technical competence in thermal drives and their growth strategy of high-volume production cannot be taken any further, beyond the current boom. The general conditions for mobility and trends in demand for mobility technology are changing rapidly. Population growth, urban densification, and ever more evident symptoms of stress from urban density, such as scarcity of living space, bottlenecks in stationary and flowing traffic, poor traffic safety and emission problems, massively aggravated by a growing demand for mobility, call for new mobility concepts. This is especially true in the future mobility growth markets of Asia and Latin America. Basically, these markets will have to build on low-emission drives and increased efficiency in using products and infrastructures.

The younger generation of users, intrigued by the "sharing economy" philosophy of "using instead of possessing", no longer insists on owning cars, a concept that is rather inelegant economically. Instead it expects reliable, flexible and cost-efficient access to modern traffic systems. Quite understandably, people want to be online and connected while traveling, too. This development will revolutionize the productive powers and production conditions of the mobility industry. The entrepreneurs of the digital-sharing economy are rapidly developing new forums, networks and applications for planning routes, optimizing traffic flows, finding parking space and sharing cars, bikes and rides. And this is leading—albeit to a lesser degree—to the emergence of digital markets for interconnected and intermodal mobility.

4   Risks and Challenges

Digitalization brings big opportunities. But it also brings risks and problems, some of which are already clearly visible today.

4.1 Legal Dimensions

"Informational self-determination", as it is known, is the individual's right to decide about the exposure and use of his or her personal data. According to the German Federal Constitutional Court it represents a fundamental right to the protection of privacy, not explicitly covered by the Constitution of the Federal Republic of Germany. Personal data is, however, protected under Article 8 of the European Charter of Fundamental Rights.

Large-scale digitalization poses a twofold threat to this right. On one hand, as a natural consequence of the way in which we use digital systems, devices and services in all areas of life, IT providers and operators collect, recombine, exploit and remarket large amounts of our personal data. So far this has not been against the law, most of the time. On the other hand cyber criminals, as well as governmental and private intelligence organizations, spy on our personal data for the purposes of surveillance and manipulation.

The digitalization of mobility and its infrastructures is one of the major areas (other than healthcare) where it will be feasible to tap into large quantities of high-quality, detailed personal data. Interconnecting vehicles and users by integrating their smartphone-based intermodal mobility assistants into digital traffic system architectures will makes it possible to keep a close track of all road users. Location-related activity patterns, combined with other data (e.g., payment processes, communication, physical data), create enormous transparency about individuals. This transparency is significant from the perspective of protecting the right to informational self-determination. Further digital penetration of the mobility market is therefore an enormous challenge. Apart from improving prevention mechanisms for critical infrastructures and international protection standards against illegal data collection, much greater awareness of the problem on the part of individual road users may be the only way to meet this challenge going forward.

4.2 Resilience

"All wheels stand still when the hacker wills it." This variant of the old slogan of the German workers' movement may become a leitmotiv of our digital future. With breakneck speed and in various disguises, information and communication technologies are invading all areas of life. Initially they come with many advantages; indeed, one can hardly predict what further conveniences and improvements they may yet bring. But the more digital and interconnected the world becomes, even in the smallest areas of everyday life, and the stronger the Internet of Things grows, the more vulnerable the critical infrastructures and our daily procedures and processes become. According to this simple rule, the greater the system complexity, the greater the risk potential—be it from envious IT specialists or, less probably but more catastrophically, from natural events. In the worst case, the consequences can have a painfully tangible domino effect on the real, digital and mixed spheres. The vulnerability of the complex and often technologically mixed, overlapping infrastructures determines exactly the vulnerability of the society built on these infrastructures as a whole. There are major threats both for the structure of public authorities and for IT-based business processes in companies and public bodies, industrial facilities, and energy, utility and traffic systems.

Let us suppose that a computer specialist is able to use malware to access the remote diagnosis servers of major automakers, for instance. By pressing a single button, he or she can manipulate a whole fleet of vehicles, bringing them to a halt or taking control of them in some other way. As the future of automotive technology lies in the electrification, digitalization and automation of functions that were previously realized mechanically, the risk is becoming more acute with each new vehicle generation. The same applies to the navigation architectures of modern global traffic, be it by sea, air or land, where the navigation devices installed in millions of passenger vehicles and trucks are vulnerable to harmful interference. It also applies to the control centers of public transit providers, the complex control and safety structures of railroads, and the road-traffic guidance systems in urban areas.

A good example of the vulnerability of close-meshed global transportation and logistics machinery to natural events were the 2010 eruptions of Eyjafjallajökull, a small volcano that no-one had previously taken much notice of. With its sharp-edged ash particles posing a serious threat to airplane turbines, it brought European flight traffic almost to a complete standstill for several days in a row. Although companies found that many business trips could be replaced by videoconferences, it also became clear that the "just-in-time" logistics of key components was so susceptible to interference that it brought European industry to the brink of a major production crisis.

Wherever the move is made from single vehicles with mostly mechanical functions to the interconnected, automated and digital texture of a highly integrated comprehensive traffic system, new risks arise that will have to be taken into account in the future. The criterion of resilience—that is, the robustness of systems and vehicles against accidental or deliberate breakdowns—will be as important for designing sustainable mobility as the criteria of environmental soundness and traffic safety.

4.3 Resource Intensity

However simple, transparent and ubiquitous digital technologies and services may appear in everyday life, the amount of resources and energy necessary to provide and operate them is huge. Equipment such as smartphones, tablets, laptops and desktop computers require rare, expensive raw materials, the mining of which is often extremely harmful to the environment. Guidance infrastructures have to be constructed, and larger and larger server farms have to be built, operated and cooled, at great material and energy costs. The still fledgling scientific and public debate about the impact on resources is sometimes known as *green IT*. Most research focuses on the energy consumed by information and communication technologies (ICT); insights into the environmental impact of the consumption, production, transportation and disposal of material resources are rare. Due to the fragmentary data, the total environmental impact can only be rudimentarily estimated, with a primary focus on energy consumption (Borderstep/IZT 2012: 41).

Some conclusions are possible about the power consumption of the ICT sector in Germany, and these are indicative of the tendencies seen in resource use overall. In most of the available studies, the following systematization is used to distinguish between different areas of consumption:

- ICT terminals in households, companies and public facilities. These include desktop computers, monitors, printers, copiers, television sets, audio and telephones
- Servers and data centers: applications, memory, communication
- Networks: Network access and core networks with network components, such as routers, switches, transreceivers, antennas, etc.; network infrastructure for land-line and mobile telephony (Borderstep/IZT 2012: 53)

What are the current and estimated future power consumption levels for these areas?

- At 33 TWh or 60 percent, terminals in private homes account for the major chunk of total ICT-related power consumption. A one-quarter increase to nearly 40 TWh by 2020 is expected, with desktop computers and TV sets being the main consumers. Companies and public facilities account for 6.8 TWh or 12 percent of total consumption. Here, it is estimated that the number of workplace computers will increase from 26.5 million in 2010 to 37.5 million in 2020. It would be possible, however, to reduce resource consumption by increasing the use of energy-efficient devices such as mini PCs, laptops, and thin clients (ibid.: 91).
- Servers and data centers consume 9.1 TWh of power. The trend toward growing power consumption in this area has been stopped by means of efficiency improvements. However, these improvements have been offset by the growing number of servers, the increasing need for memory, and more network technology (ibid. 2012: 44).
- It has not been possible to reduce material consumption. The share of electronics, in particular, continues to rise.
- For network access and core networks, power consumption was estimated at 6.4 TWh in 2007 (Fraunhofer IZM/ISI 2009: 13). This includes telephone and Internet network access, and the core network itself with its network components. A quadrupling of power consumption from 8 TWh in 2010 to more than 32 TWh in 2020 is expected. The network sector is therefore of

particular importance when it comes to resource politics (Borderstep/IZT 2012: 91).

Suppose older equipment is replaced by newer, energy-efficient equipment. Is there a chance that the improvements in energy efficiency would then offset the environmental impact of producing and distributing the latter and disposing of the former? A study commissioned by the German Federal Environmental Agency (Öko-Institut/Fraunhofer IZM 2012) examined this question—known as "energy amortization"—for laptops, asking how much more efficient a new laptop would have to be in order to justify replacing an older and less energy-efficient one? It was found that production is responsible for 56 percent (or 214 kg of CO_2 over 5 years) of the total greenhouse gas emissions, which is more than what is emitted during the equipment's useful life. The environmental cost of production is so high that, in realistic periods of time, it cannot be offset by a more energy-efficient lifetime operation. If replacing an older device with a newer version improves energy efficiency by 10 percent, amortization will take 33–89 years. It is therefore crucial to enhance the lifespan of new equipment in order to reduce the environmental burden created by the production phase.

It is thought that, due to fast-growing total demand, ICT-related resource consumption will continue to rise in the future. While there is room for significant efficiency improvements in both end equipment (e.g., through miniaturization) and data servers and networks, we may expect to see an increase in ICT-related power consumption in Germany from about 59.6 TWh in 2010 to more than 90 TWh in 2020. This would amount to a 50 percent increase, bringing the share of ICT in total power consumption in Germany to nearly 20 percent. This could put considerable pressure on the power grids, with consequences for security of supply and the environment (Borderstep/IZT 2012: 77).

Globally, about 2.5 billion people have access to the Internet. In 2017 we will reach a point where half of world's population—or 3.6 billion of the then 7.2 billion people in the world—will be online (Greenpeace 2014). As Internet and cloud usage intensifies, global power consumption will also grow disproportionately strongly: Increases of 60+ percent by 2020 are to be expected due to the growing online population and its dependence on the Internet. User data will be stored in server parks consuming large quantities of energy. High-performance processors use a great deal of energy and must also be permanently cooled, right around the clock. Global cloud computing already consumes more power than the whole of Germany. If the Internet were a country it would have the sixth-biggest power consumption globally.

Another problem is that the Internet's environmental footprint is largely concentrated in places where energy is produced in a particularly dirty way. The Greenpeace study quoted above (Greenpeace, ibid.) examined 19 globally leading IT businesses that were vigorously promoting the shift toward cloud computing and which themselves process a major part of the data on the Internet. Six major cloud brand leaders (Apple, Box, Facebook, Google, Rackspace, Salesforce) were found to have committed themselves to the goal of using only sustainable energy to power their data-processing centers. For the time being, however, only Apply relies exclusively on renewable energy (solar, hydroelectric and geo-thermal) to fuel its server park in the southern United States. Together with Google and Facebook, the company has persuaded the largest US power supplier, Duke Energy, to open up its market to green power. Companies that are doing badly, by contrast, include eBay and in particular Amazon, which is described in the report as "one of the dirtiest and least transparent companies in the Internet." Among Germany companies, the study looked at IBM's data center in Ehningen (Baden-Württemberg), finding that it obtains 22 percent of its power from renewable sources, 14 percent from gas, 18 percent from nuclear energy, and 45 percent from coal.

Because of the great need for cooling, more and more data centers are being built in Scandinavia. In northern Sweden, for instance, Facebook has set up a data center whose servers are cooled with outdoor air. Data centers are also being planned in Iceland and Norway, as well as in Finland, where Google has put into operation a data center cooled with seawater. This trend can be expected to continue. IT companies are becoming important players, able to make major contributions to the increased use of renewable energy.

In the coming decades, outside the United States the focus will be on China, where a major part of Internet growth will take place. If companies in China rely on dirty energy to build and expand their Internet infrastructure, this could have disastrous consequences for carbon emission and air pollution. Innovative approaches, however, could make a sizeable difference.

4.4 Rebound Effects

Closely linked to the question of resource intensity is the question of "rebound effects". Technological innovations are commonly used to minimize the expenditure of time, capital and resources. However, these technological efficiency improvements are linked to increased expenditure elsewhere, offsetting the initial savings—a problem that is increasingly becoming the focus of economic debate.

The term *rebound* refers to this effect. In a high-profile study dealing with this problem, Madlener and Alcott (2011) suggest the following definitions. The term *rebound* covers all impacts on the demand situation in an economy resulting from a certain improvement in technological efficiency—not just the direct impact on the products (goods and services) that the technological advance has made more efficient. For instance, many studies examine consumers' behavior after buying a more fuel-efficient vehicle, such as the number of miles driven or perhaps the purchase of an additional vehicle. Other studies measure how much more people heat their homes after they have improved the thermal insulation of the property and heating has become cheaper. This type of rebound is called a "direct rebound". "Indirect rebounds" are all other types of effect. After an efficiency improvement, for instance, consumers are left with greater purchasing power that they can use to buy various goods and services. Furthermore, energy itself becomes cheaper, as the efficiency enhancement leads to a (temporary) reduction in demand; this in turn boosts demand.

The various innovation paths described in Sect. 3 aim at increasing the efficiency of using certain products or the system as a whole. Since digital innovation itself often comes with significant efficiency enhancements, rebound effects must be taken into account in the future if the use of digital technologies is to serve the purpose of overall environmental optimization.

One immediate conclusion relating to digital rebound effects is that, if our ultimate aim is to achieve ecological ends, digitally supported traffic system innovations should always be implemented as part of an overriding target system and, if appropriate, supported by complementary actions. In this way of thinking, any optimization of the urban traffic flow for moving and stationary vehicles by telematically guiding digitally interconnected fleets of vehicles would necessarily involve regulating and limiting the resulting possible increase in automobility by means of appropriate fiscal and regulatory instruments.

Finally, let us turn to the possible more general rebound effects of digitalization on society. Here, a metaphor may be helpful. A society that is functionally highly differentiated and depends for its integration on a high degree of mobility—with roads, pathways and exchange processes—is like a biological organism with its blood vessels, nerve tracts and control centers. It is to be feared that the digitalization of the overall societal and economic organism may have an effect similar to that of several liters of caffeinated beverages or stimulating drugs on the human organism, namely a speeding-up of societal processes and short-term increases rather than sustainable, permanent developments. This will have all the inevitable consequences, such as increased resource consumption, greater use of space, the destruction of existing social structures and institutions, and so on. Of course, this would have little in common with the goal of continuous development aimed at sustainability. Indeed, the only way to offset it would be to tax the increased resource consumption resulting from the efficiency enhancements, as suggested by v. Weizsäcker et al. (2010: 303).

5   From Big Oil to Big Data: Solar-Digital Mobility

I personally believe that digitalization has the power to turn the non-sustainable mobility of the oil era into sustainable system innovations, creating what we may call a "solar-digital era". We can potentially make mobility sustainable without loss of quality as regards the general accessibility of the infrastructures we need to support everyday life. How? By combining digital technologies with post-fossil and ultimately *solar* propulsion technologies. By massively improving the efficiency of how we use vehicles and infrastructures in passenger and goods traffic. By making comprehensive efforts to reduce traffic with the help of digital virtualization technologies. And by developing innovative spatial planning instruments for promoting densification and regionalization.

This will only work, however, if there is a strong political commitment to wisely promoting the vision of a new intermodal mobility. Moreover, this must be implemented in a way that takes into account the downsides of otherwise extremely helpful digital technologies.

As far as resources are concerned, we have to consider whether the use of digital technology could actually make the burden of transportation services on the environment even heavier. We still do not know whether the added digital effort involved is offset by the resulting efficiency optimization, ultimately resulting in real savings. What is absolutely necessary are closed resource cycles, which in turn require new types of industrial development and production. The question of—possibly much lower—future employment is one of the biggest challenges that we will face.

In conclusion, however, it is worth stressing again that what is needed is great political courage and the ability to acknowledge that only action early on, even going against the short-term interests of established players, can prevent our society from heading into a dead end—from where we would be acting with our backs up against the wall.

References

Borderstep/IZT (2012) "Green IT—Nachhaltigkeit", Borderstep Institute for Innovation and Sustainability, Institute for Future Studies and Technology Assessment (IZT). Report for the Commission of Enquiry on "the internet and the digital society" of the German Bundestag, committee bulletin 17(24)058, Berlin

German Federal Environmental Agency (2012) Entwicklung der spezifischen Kohlendioxid-Emissionen des deutschen Strommix 1990–2010 und erste Abschätzungen 2011. Dessau-Roßlau, Apr 2012

Greenpeace (2014) Clicking clean—wie Unternehmen ein umweltfreundliches Internet erschaffen. Abstract, Apr, Hamburg

ISI/IZT (2009) Rohstoffbedarf für Zukunftstechnologien, Einfluss des branchenspezifischen Rohstoffverbrauchs in rohstoffintensiven Zukunftstechnologien auf die zukünftige Rohstoffnachfrage. Fraunhofer Institute for Systems und Innovation Research (ISI), Institute for Future Studies and Technology Assessment (IZT), study commissioned by the Federal Ministry of Economy and Technology, Berlin

Madlener R, Alcott B (2011) Herausforderungen für eine technisch-ökonomische Entkoppelung von Naturverbrauch und Wirtschaftswachstum unter besonderer Berücksichtigung der Systematisierung von Rebound-Effekten und Problemverschiebungen. Commissioned by the Commission of Enquiry on "growth, prosperity, and quality of life" of the German Bundestag, Zurich

Öko-Institut/Fraunhofer IZM (2012) Siddharth Prakash, Ran Liu, Karsten Schischke, and Lutz Stobbe. Zeitlich optimierter Ersatz eines Notebooks unter ökologischen Gesichtspunkten. Commissioned by the German Federal Environmental Agency, text 44/2012, environmental research plan of the Federal Ministry of Environment, Nature Conservation, and Nuclear Safety, FKZ 363 01 322; UBA-FB 001666, Berlin

v. Weizsäcker, E et al (2010) Faktor fünf—die Formel für nachhaltiges Wachstum. Munich

Hitachi Energy Trading Optimizer

Markus Hartwig and Olaf Heil

1 Introduction

Worldwide the energy supply is being reshaped. The provision of energy based on renewable and sustainable resources is changing value creation of power markets worldwide. Globally 286 billion $ were invested into renewable energy sources in 2015 which is more than twice as much as for conventional thermal power that very same year.

Germany and the European Union, due to its political agenda is in the fore front in changing the supply of electricity from a centralized system, based on conventional fossil-fired power plants and nuclear, to a decentralized system based on renewable energy sources such as on- and off-shore wind, hydro power, bio-fuels and solar.

In 2015 supply from renewable power in Germany reached 36% of total demand, an all-time high up to this point. This makes power supply ever more weather dependent. For times when renewable power plants cannot supply enough energy conventional supply is still required as backup solution. In Germany the conventional power plant portfolio contains basically nuclear (to be shut-down 2022), lignite, hard coal and gas. As weather is rather unpredictable these assets need to be highly flexible in order to have minimal lead times with respect to capacity changes.

Renewable power has fundamentally changed the market dynamics within the European energy space. In times of ample amounts of wind and sun already today Germany can fully cover its demand from renewable sources. Under these circumstances conventional power plant is no longer required. Unfortunately, lignite power as much as nuclear power cannot be shut down temporarily on a short term basis which means there is significant over supply in the market. This energy

M. Hartwig (✉) • O. Heil
Hitachi Europe GmbH, Düsseldorf, Germany
e-mail: hartwig@mcg.eu.com; olaf.heil@hitachi-eu.com

© Springer International Publishing AG 2017
T. Osburg, C. Lohrmann (eds.), *Sustainability in a Digital World*, CSR,
Sustainability, Ethics & Governance, DOI 10.1007/978-3-319-54603-2_14

surplus leads to very low power price level at the stock markets, even in over 150 h per year negative electricity prices. As a result, a significant number of assets are no longer competitive and in consequence were decommissioned. The German utilities suffer by this development dramatically.

In future prices will remain on lower levels, but also get more and more volatile as renewables increase further. The same time pricing and power trading becomes much more complex and as a result the price forecasting gets the key criteria. Big-data based, AI supported analytics for power trading and IoT based power generation and demand steering will be the way out for customers.

2 Market Player

The price development has long lasting effects on the market players and their respective investment road map. The overall target of the Energiewende is to build a low emitting and finally a CO_2 free economy. In a first step the power plant portfolio should be contained of renewable and gas based power supply. This is why in the period of 2005–2010 a significant investment took place to scale up and renew natural gas based power generation. These investments were based on an electricity price level of 55 €/MWh and hourly prices being beyond 100 €/MWh in over 150 h per year: It was common market believe that power prices would continue to rise.

As of today power prices are more in the dimension of 25 €/MWh with the most expensive hour in 2015 having been 70€. Clearly this has impacted the profitability of power plants in many ways. As a result, power plant operators not only need to find each and every profitable hour of the year but also need to capitalize on the power plants' inherent flexibilities in order to sustain profitability in the long run.

3 New Market Participants

The market design for renewable power has undergone three steps, from guaranteed feed in tariffs to the optionality of trading renewable power on the wholesale market to a must trade renewable power on the wholesale market. The existing market design in combination with the legal framework makes it paramount duty that the market participants have a profound understanding of the short term market dynamics in order to optimize their profit and loss in an increasingly competitive market environment as they need to refocus their value proposition form subsidy system participation to true market competition.

To sum it all up incumbents and new entrants struggle to come to terms with the market dynamics that are due today. As a result, market participants that own 117,000 MW in generation capacity are in need to find tools that can help them to adapt their value proposition to the new market environment. The Hitachi Energy Trading Optimizer (HETO) is a first successful approach to provide such a service.

4 Hitachi Energy Trading Optimizer

Hitachi Europe has defined Smart Energy and Industry 4.0 as one of their future growth areas. Therefore we defined Stadtwerke and Industrial Clients in Germany as the nucleus for growth. The platform of the Hitachi Energy Optimizer will be consisting of variety of applications covering the entire value chain of the energy business. It will be developed in different phases and will be rolled into the market space through different releases.

In its first version HETO will permit the customer to optimize its short term power positions, in particular on the generation side. The tool gives recommendations as how to optimize the generation assets within the short term time horizon. The scope of the product's first version is tailored to supporting the different players in capturing value in the market space that is driven by renewable power, as described above.

The software comprises different components that are laid out in the following outline (see below Fig. 1). In principal the software contains three basic components:

1. Customer specific elements: Each customer has a unique supply and demand portfolio. This information needs to be uploaded into a platform which is

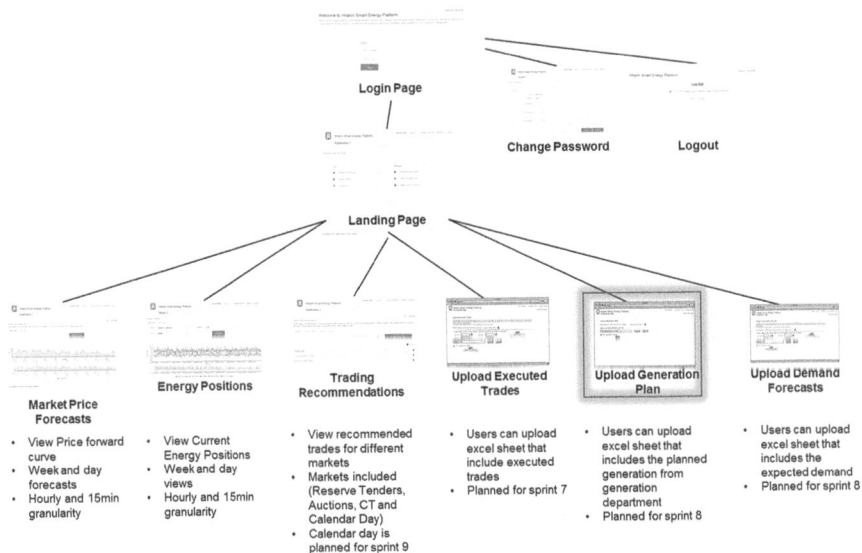

Fig. 1 Software design outline for generation assets

undertaken through specific user interfaces. Given the customer information the software gives recommendations as to how the customer portfolio shall be optimized for value optimization. The different user interfaces will be described in greater detail further down in this paper.

2. Market information: In order to find the optimal trading strategy market and other kind of data needs to be made available in real time. Therefore, the software contains different interfaces that permit an update of market information.

3. Calculation kernel: The calculation kernel takes the details of the respective customer portfolio and calculates the optimal trading strategy for all products at the power market.

Generation Plan

A generation portfolio of a Stadtwerk or an eligible industrial client can contain many different types of assets, with varying amounts of flexibility. Simple assets such as a Biogas plants may be relatively simple to flex up and down, however more complex assets such as a heat-driven Combined Heat & Power plant may consist of multiple interdependent components that must be controlled separately. One of the longer term functional goals for the product is to be able to engineer and maximize flexibility by smartly modelling power plants.

Using the available flex, the product will schedule generation against predicted market prices in order to maximize profitability. When creating optimized schedules, the product needs to respect the technical constraints of different assets, for example ramping speeds, minimum run times and must-run periods (see below Fig. 2).

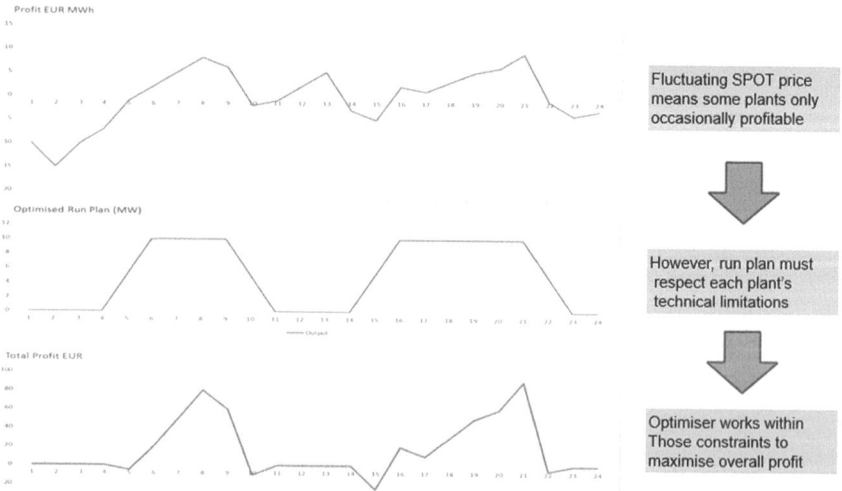

Fig. 2 Interaction of profits, technological constraints and predictes market prices

The generation schedule interface allows the platform user to upload the details of the actual planned generation for a specific period. The data represents the expected power plant output as planned by the generation team. The user can upload as many plans as they need and the latest plan will overwrite the existing data. The granularity of the data upload is 15 min.

Demand Forecast
The customer's demand forecast is a key input to HETO, it is required by the trade optimization module. For the current phase the data will be uploaded manually.

Executed Trades
A key input to HETO is the customer's current energy trading positions for each 15 min period. The product will provide a platform that will allow customers to connect their back-office systems (e.g. trading systems) so that this information can be uploaded automatically, rather than relying on manual data input.

Trading Recommendations
The Trading Plan Page is the one of the interfaces in the platform. It allows the platform user to view the recommended trades by the platform for different markets.

Once the assets have been scheduled optimally, the total energy output needs to be combined with other energy inputs & outputs to create aggregate energy positions for each 15-min period. These positions are then traded until balance is achieved (supply = demand) for each 15-min period (see Fig. 3 below).

As there are many different markets energy can be traded on, the product has to optimize the combination of all markets to meet the joint objectives of clearing the aggregated energy positions for each 15-min period and maximizing profit.

Optimization Kernel
The optimization kernel is built within a data science lab that uses the latest in terms of optimization technology:

1. Multivariate regressions: Multiple linear regression is a generalization of linear regression by considering more than one independent variable, and a specific case of general linear models formed by restricting the number of dependent variables to one.

Fig. 3 Interface to view the recommended trades from the platform

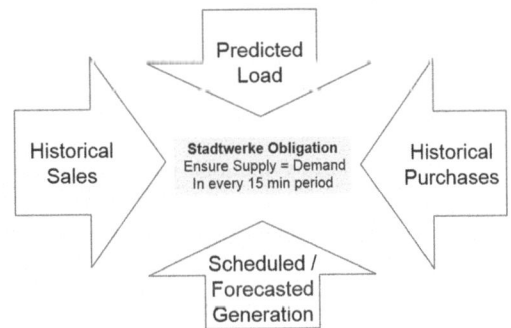

2. Neural networks: In machine learning and cognitive science, artificial neural networks (ANNs) is a network inspired by biological neural networks (the central nervous systems of animals, in particular the brain) which are used in the next HETO version to estimate or approximate functions that can depend on a large number of inputs that are generally unknown. Artificial neural networks are typically specified using three things:

   (a) Architecture specifies what variables are involved in the network and their topological relationships—for example the variables involved in a neural network might be the weights of the connections between the neurons, along with activities of the neurons.
   (b) Activity Rule: Most neural network models have short time-scale dynamics: local rules define how the activities of the neurons change in response to each other. Typically the activity rule depends on the weights (the parameters) in the network.
   (c) Learning Rule: The learning rule specifies the way in which the neural network's weights change with time. This learning is usually viewed as taking place on a longer time scale than the time scale of the dynamics under the activity rule. Usually the learning rule will depend on the activities of the neurons. It may also depend on the values of the target values supplied by a teacher and on the current value of the weights.

3. Multivariate stochastic optimisation for non-linear asset optimisation: In the field of mathematical optimization, stochastic programming is a framework for modeling optimization problems that involve uncertainty. Whereas deterministic optimization problems are formulated with known parameters, real world problems almost invariably include some unknown parameters. When the parameters are known only within certain bounds, one approach to tackling such problems is called robust optimization. Here the goal is to find a solution which is feasible for all such data and optimal in some sense. Stochastic programming models are similar in style but take advantage of the fact that probability distributions governing the data are known or can be estimated. The goal here is to find some policy that is feasible for all (or almost all) the possible data instances and maximizes the expectation of some function of the decisions and the random variables. More generally, such models are formulated, solved analytically or numerically, and analyzed in order to provide useful information to a decision-maker.

4. Factorial Markov modelling for complex asset steering: A hidden Markov model (HMM) is a statistical Markov model in which the system being modeled is assumed to be a Markov process with unobserved (hidden) states.

In simpler Markov models, the state is directly visible to the observer, and therefore the state transition probabilities are the only parameters. In a hidden Markov model, the state is not directly visible, but the output, dependent on the state, is visible. Each state has a probability distribution over the possible output tokens. Therefore, the sequence of tokens generated by an HMM gives some information about the sequence of states. The adjective 'hidden' refers to the state sequence through which the model passes, not to the parameters of the model; the model is still referred to as a 'hidden' Markov model even if these parameters are known exactly.

Hidden Markov models are especially known for their application in temporal pattern recognition such as, muscular score following partial discharges and bioinformatics.

5 Conclusion

As the energy business landscape in Germany had a dramatic change over the last two decades by having free tradable power markets, free choice of supplier, the increase of renewables and the shut-down of nuclear power plants, the market player have to meet lots of new requirements and challenges; a transformation of the energy system that is expected to continue. Hitachi Europe builds IoT-solutions, based on big-data analytics and artificial intelligence systems for the energy business to help them to overcome the issues. As these disruptive changes will happen in several countries over the next years, the solution will be multiplied globally in future.

Digital Sustainability in the Banking and Finance Sector

Görkem Çokçetin

1 Introduction

Banking has been one of the well-structured and regulated businesses in the corporate business world. The main asset of banking is the intermediate role between lender and borrower with the trust built on banks corporate presence and government license.

In the last quarter of twentieth century, moving funds between customers and countries start to become more important business for banks due to the expansion of the electronic money. Payments, cash management and foreign trade grew and became the engine of moving funds. Billons of Euros and Dollars started to be transferred move from one company to other and one country to another within hours' time.

Also banking products especially investment products get more complex in this era. Understanding the logic and calculation of such products become so complex for ordinary customer without special calculations tools.

During this era and maybe till the end of first decade of 2000s finance sector has been using the latest and strongest IT systems and infrastructures and heading technology for the customers. These systems served banks well until the past decade, when the IT environment changed markedly, and Web communications, network computing, and plug-and-play system design emerged as keystones of high-performing IT platforms (Heidmann 2010).

In the first years of internet, banks were still holding their leading position and offering services through Internet to their customers. The first web-based services were mostly informative such as an introduction of the bank, branch lists and operating hours and daily exchange rates.

G. Çokçetin (✉)
Yapı ve Kredi Bankası A.Ş, Istanbul, Turkey
e-mail: gorkem.cokcetin@yapikredi.com.tr

© Springer International Publishing AG 2017
T. Osburg, C. Lohrmann (eds.), *Sustainability in a Digital World*, CSR,
Sustainability, Ethics & Governance, DOI 10.1007/978-3-319-54603-2_15

Then checking account and credit balance and transactions history in the customer's account started to be reachable for the customers who are lucky to have both internet access and internet banking password. As internet access prevalent through land telephone lines (dial-up) more services started to be useable for the customers.

The main reasons for offering new services to customers and promoting internet banking were mostly for cost reduction and profitability. Banks were transferring their load in costly and mostly less profitable transactions such as internal money transfers (EFT) and bill payments to customer thanks to internet banking. In these period only back and middle offices have used and benefited from digital technologies in order to save money while front offices have been labor-intensive and relationship-based.

In the early 2000s more services became available in internet banking also with the demands coming from customer side. Now customers were defining the services which they want to access in the internet but the interface were still define and design base on the bank core systems.

As the smart phones appeared in the mass market and new mobile operating systems such as android, IOS were introduce to the final users, banks leading and conducting position become questionable.

Customers now have the latest technologic devices and the operating system is very different from the core banking system. Most of the repeating transactions are being done automatically by the bank's core system with a few needs to human interactions. This part of the story is not very new but the new part is the orders which triggered these transactions which were taken from customers by employees and insert in the system is now being done by the clients on-line from internet or mobile banking. Also information about the balance of an account or transactions list is available for the clients in digital format without a need to go or call the branch.

Banks and other financial institution have found themselves in a position to develop dedicated interfaces to each operating system and to each type of device. Even if the operating systems are same in two devices like a smart phone and a tablet, the device type and version must take into consideration as to give the same look and feel for the customers regardless of the device they are using.

2 Digital Treats for Banking

Since the commercialization of digital photography more photos were taken from the rest of the history. But Kodak, the biggest photo film producer who also invented the digital camera was bankrupt. Kodak was bankrupted because the top management of Kodak saw digital photography as a treat to their well going daily business and not invested on the digital technology.

However Instagram, as a social photography sharing app which I has a few hundreds of employees, market value is higher than a lot of chemical company with thousands of workers.

Of-course the banks and the other financial institution will not be vanished due to digitalization in the near future. People will want their money to be safer than their selfies (Chris Skinner-BKM 2016). Regulations and the licenses and the governments will support the banks for long years.

After the Global Financial Crisis in 2008, the financial technology (Fin-Tech) sector has become one of the fastest growing entrepreneurial areas in the world for start-ups that aim to disrupt traditional banks (Melike Belli-BKM 2016).

Digital Fin-techs start offering the finance sectors clients faster and almost free services from payments to wealth management. Big technology firms like Apple, Samsung are trying to spread their payment system into different markets. Fin-techs and technology firms become both rival and partner in the same time.

The most valuable assets of banking in the digital world are customer base and the customer data they have in their system. The banks and other financial institutions must decide whether to adopt their selves to digital era and create new services and apps like Instagram which attract the digital age customers with almost free or lost their profitability and effects day by day.

As a result, an opportunity for banks and digital Fin-Tech companies has arisen as welcoming innovation together in cooperation. This will help both of them to increase the customer satisfaction and provide them sustainability with corporate advantage.

3 Alternative Models for Sustainability

Everyday more and more process and services in the back and middle offices in banks and financial institutions become fully digital and automatic. This means employees in the roles will lose their jobs or in will be shift to the front offices.

Banks and financial institution have to find new models by using the digital technologies for their own sustainability but also sustainability of a reliable finance system. In the coming pages, I want to share the three alternative paths or models for the sustainability of banking/finance sector.

These models include finding new businesses and customer for the employees shift from back office to front office and using digital assisted services for new value and revenue generation.

Banking the Unbanked and Underbanked
Even in mature economies like US there is a high percentage unbanked and underbanked population. According FDIC survey figures approximately 33% of adults age 18+ are underbanked or unbanked in US (FDIC 2014) Notably, this

represents an increase from 2008, which suggests that more and more people are either unserved by traditional banking institutions or are finding competitive alternatives to traditional banking.

Also less developed parts of the world like Africa and Asia the portion of the unbanked and underbanked population is higher than the banked population. However there is a significant change from 2011 and 2014; in 3 years 700 million people became account holders at banks, other financial institutions or mobile money service providers, and the number of "unbanked" individuals dropped 20% to 2 billion adults (Global Findex 2014).

Technology in the forms of telecommunication and digitalization played an important role in these improvements. One way to rapidly expand financial inclusion is new technology, particularly mobile money accounts. MPesa mobile phone-based money transfer, financing and microfinancing service, launched in 2007 by Vodafone for Safaricom and Vodacom, the largest mobile network operators in Kenya and Tanzania played a key role in this financial inclusions. In Sub-Saharan Africa 12% of adults have a mobile money account. In 13 countries, usage exceeds 10% and among those, Cote d'Ivoire, Somalia, Tanzania, Uganda, and Zimbabwe more adults using a mobile money account than an account at a financial institution (Global Findex 2014).

As digitization eases the inclusion financial institutions must find new ways to reach and serve this population in a wider perspective for their sustainability. Small agents like branches in remote districts or less developed part of the cities may help further inclusion and products usages. Even basic transaction could be operating from digital channels human involvement is essential in advisory or credit evaluation like cases.

A relationship manager, without a dedicated branch, equipped with digital tools (pads, smartphone etc.) which are able to access bank's core system for monitoring the accounts, sending customer orders and doing fast credit evaluation will may be good solution. These RM's reach small business owners in the less developed parts of the cities and helped them for the financial inclusion. This model is being in recent years in Yapı Kredi—Turkey for new SME customer acquisitions.

Mobile branches of RBI in India to serve the inhabitants of remote villages and floating bank branches of PT Bank Rakyat in Indonesia are two very successful models. PT Bank Rakyat customized four boats, complete with ATMs, in order to better serve millions of Indonesians who still have no bank account.

Supporting the Sharing Economy

If you take definition of the "Economic problem" as satisfying the unlimited human needs, wants with limited resources, sharing economy is becoming a good solution.

Sharing economy is on the rise with digitalization. Thanks to the digital abilities of our apps and smart phones, utilizing idle resource becoming easier. These services are possible due to a global digital distribution, which creates another challenge for incumbents in traditional industries on an infrastructure level.

The sharing economy is expected to generate revenues up to $335 billion by 2025, and its impact is predicted to affect nearly all industries (pwc Blogs 2014).

Uber and Airbnb are already a new paradigm for finance. Rather than buying cars, hotel rooms and more, bank customers rent time and space on-demand through apps like Airbnb and Uber or Lynt. That means less demand car loans and less cash flow in your hotel customer.

There are more sharing economy practices less known then Uber or Airbnb but still working and creating benefit for communities. You can even find someone to take care of your dog when you are in vacation with a less amount you will be paying to a "dog hotel".

So how banks and financial institutions can integrate their selves in the sharing economy for their sustainability? The easiest and fastest answer is getting their share from payments. Even though P2P payments apps are getting more popular banks still have the upper hand. Because most of the property/god owner participants of these apps are already have an account and credit card with banks. Turning them to POS merchants and small business owners is the key.

By mobile and virtual POSs and digital wallets Banks can easily beat other non-bank P2P payment rivals. But to beat them, banks must provide these services in a cheaper and more flexible way than they offer now.

Also giving small revolving credit limits to property owner to buy a new car or to improve the conditions in the rented flats based on the transaction from Uber or Airbnb will integrate the banks to sharing economy.

There are already some non-bank companies offering services in this concept. Square is most successful example. After downloading the Square app with a small gadget you plugged in to the smartphone everyone become a merchant. Square transfers the amounts to customer account in any bank with cutting a fix rate. Square also offers sellers financial and marketing services, including small business financing and customer engagement tools (Squareup.com).

Square become so successful that acquires more than a million customers in first 3–4 years. Squared initial public stock offered in Wall-Street in December 2015. So banks and FI's must quickly find ways to integrate their selves in to this part sharing economy in the near future.

Another sharing economy model banks and FI's must involve in digital world is crowdfunding. Crowdfunding is a viable option in digital world when you do not want to take out a small loan but would like some extra cash for a variety of reason.

With crowdfunding, donors contribute to cause to help, idea or the person to reach the goal. It falls between peer-to-peer loans, which borrower repay with interest, and fundraising sites, where borrower do not offer any compensation.

Personal crowdfunding is different than business crowdfunding in that you do not offer stocks or equity in a business as repayment. Personal crowdfunding could be define as raising funds that barrower do not repay but instead offer rewards to incentivize people to donate and to thank them for their contributions. Some of these sites include Kickstarter, Indiegogo and Tilt.

You may think how banks or financial institution can involve in a model where no repayment or interest expected. Even though there are no repayments or interest paid banks could support funded projects or people with their services and experience.

Banks could involve in payment-processing for fee, although but require to use a service bank offers like internet banking or mobile wallets. In these services donors direct deposit and transfer funds to project's bank account.

Also experience in the different sectors and customer database could be offered as an asset to crowdfunded projects (more details in the next section). Donors could be informed about this no monetary supports of the banks to acquire new digital age customers and rewards could be offered in bank services.

Using the Bank's Big Data for the Customers

Most of this data are not structured to be use for digital analytics purpose. Big-data is a very hot topic in digital world and banks have more "big data" they think and use.

One of the most valuable assets of banks and FIs is their database about their customer. Banks were collecting many data from many different sources for many years for different reasons. Also by the digitalization and social media there are new data sources available for the banks and Fis which they can combine with their existing data about their customers.

The competitive nature in banking and financial services will assure that those that take advantage of these new data sources to augment what they know about their business will continue to be leaders. Some will likely leverage their advanced footprints to offer data subscriber networks, thereby going into competition with data aggregators and further monetizing their investments (Oracle Enterprise Architecture White Paper 2015).

It is obvious that with big data and analytics banks have more, deeper and faster insights about their customers. There are two ways of using these insights for the sustainability in digital world.

First is offering personalize offers and improved customer service to customer leading better cross sales results. This is what all the banks are mostly trying to achieve with big data in these days.

However a more innovative approach could be creating cross fertilizer digital services for the customer to reach other customers. Banks can create B2C services to conduct the personalize offers to individual customers to be met by small and mid-size business customer.

A good new example is Yapı Kredi Bank's merchant self-service campaign management service. In the service Yapı Kredi customer merchants can create its own campaign offers and communicate via mails and SMS to a customer base which banks already create based on sectoral spending prediction. Bank collects a fee regarding the size of the target group and campaign duration. The results are definitely better for the merchants which they spend more money to advertisement in web or in another channel. Also customers trust is higher as bank involved in these offers.

Similar models could be created easily especially for building direct connections from producers and small farms to customers. Small farmers could be protected and organic production could be supported with such kind of cross fertilizer services.

Financial services and banking companies gather sensitive data that in the wrong hands could lead to liability claims and worse. Securing access to the data and keeping privacy of the customers is crucial in these kinds of services. But banks surely will develop efficient and secure ways of using the customers' data for the customers and with the customers. These help banks to generate revenue from the data as a data provider also helps the society for sustainability with smarter offers.

4 Conclusion

It is obvious that banking will not lose power with the digitalization but the banks probably will. New possibilities will increase the type and volumes of transactions fin-techs and technology firms will have the higher portion than they have today.

Unless Banks and financial institution will new ways to create new labor friendly expansions they will become smaller and more digital.

Fostering innovation and entrepreneurship in the organization is essential to cope with such a transformation. Banks and financial institution must courage their employees to build new ties with digital economy firms for new business opportunities. Maybe also incubate and then accelerate the start-ups founded by their personals.

Exuberance which hopefully occur in the digital world would only be distribute smartly for sustainability with such efforts.

References

Chris Skinner-Digital Banking-BKM Pub (2016) Big data in financial services and banking— Architect's guide and reference architecture introduction. Oracle Enterprise Architecture White Paper-2015

FDIC (2014) https://www.fdic.gov/about/strategic/report/2014annualreport/index_pdf.html

Global Findex (2014) http://www.worldbank.org/en/programs/globalfindex

Heidmann M (2010) Overhauling banks' IT systems. McKinsey & Company

Melike Belli-Banking and Fintech/Developing a fintech ecosystem in Istanbul BKM (2016) http://crowdfunding-sites-review.toptenreviews.com/

http://www.wsj.com/articles/square-ipo-may-prove-turning-point-for-technology-1447981621

http://crowdfunding-sites-review.toptenreviews.com/

Oracle Enterprise Architecture White Paper (2015) Big data in financial services and banking. Architect's guide and reference architecture introduction

pwc Blogs (2014) https://www.pwc.co.uk/issues/megatrends/collisions/sharingeconomy/outlook-for-the-sharing-economy-in-the-uk-2016.html

Fintech: The Digital Transformation in the Financial Sector

Thomas F. Dapp

1 Challenges for Traditional Banks

Breathtaking. There is no word more fitting to describe the profound changes unleashed by digitalisation. Digital structural change is an evolving process—and the most compelling innovations are still in their infant stages—but its momentum is nevertheless remarkable. Digital change is unstoppable and in full flow across numerous sectors, interacting constantly with the forces of globalisation. The digital revolution is being driven by the ongoing exponential rise in data volumes, the use of microsensors and biometric recognition software, the significant increase in memory capacity, and the fact that, true to Moore's Law, processing power continues to double at frequent intervals, while prices come tumbling down (See Brynjolfsson and McAfee 2014). The hidden underlying factors of economic network effects, economies of scale and the increasing importance of peer-to-peer mechanisms play an equally fundamental role in the way in which digital technologies are virally penetrating the market (Bahr et al. 2012). These rapid changes are occurring discernibly faster than analogue innovation cycles; and frequently, and across many sectors, this acceleration is underestimated. The traditional finance sector is no exception (see below Fig. 1).

Digital structural change is having a radical impact on traditional banks. Despite the very tight squeeze on some margins, the fallout from the financial crisis, the changing consumption behaviour of customers, and the increasingly stringent regulatory requirements, banks now need to invest more in digital technologies and adapt comprehensively to the modern internet age. The main challenges lie in the need for the established banks to develop—primarily of their own accord—into primarily digital, platform-based ecosystems, while at the same time remaining

T.F. Dapp (✉)
KfW Bankengruppe, Frankfurt am Main, Germany
e-mail: thomas.dapp@kfw.de

© Springer International Publishing AG 2017
T. Osburg, C. Lohrmann (eds.), *Sustainability in a Digital World*, CSR,
Sustainability, Ethics & Governance, DOI 10.1007/978-3-319-54603-2_16

189

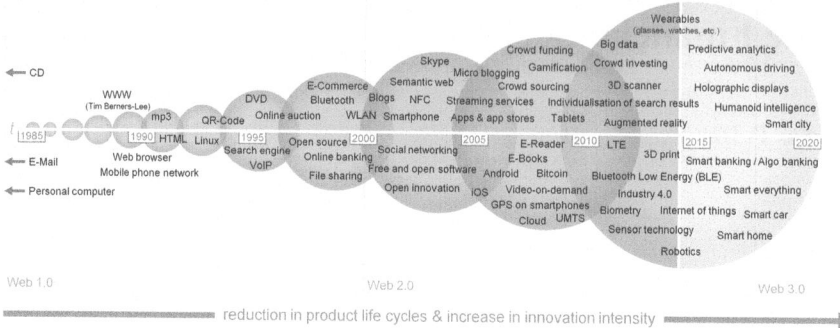

Fig. 1 Milestones in the internet age (own illustration)

open-minded towards the opportunity of entering into strategic alliances with external financial and technology service providers along their entire value chain.

At present there are signs that many companies across all sectors, including traditional banks, may be underestimating the Herculean challenge of "going digital". Initial reforms and/or innovations at traditional banks are visible. But within certain business divisions the process of adapting to the digital age is, in places, only unfolding at the customer end of the value chain, such as in online banking services for retail customers with useful web-based services; the use of biometric recognition software; or further proprietary (digital) financial services. The strategies being used and communicated continue to be driven by the silo approach, which, though traditional, is not particularly conducive to innovation.

This alone will not enable banks to achieve the success they seek. It is not sufficient to equip individual business divisions or individual sales channels with advanced internet technologies in isolation. Success will only follow from the holistic adoption of an adequate digitalisation strategy (Dapp 2015). Companies must include all of their business divisions, and they must provide suitable internal and external (preferably open) programming interfaces (application programming interfaces, APIs) for the adoption of new technologies. The impact of such changes is felt by all internal and external business areas, such as R&D, sales, service, quality management, legal and compliance, human resources, and marketing, meaning that all internal administrative and back-up processes also play a role.

The New Players in the Financial Market Come from Outside the Banking Sector

The new market players from the non-banking sector, by contrast, have an almost perfect understanding and command of the language of the internet. In response to the fast pace of innovation, digital platforms operating internationally are positioning themselves increasingly in a range of technology-driven markets. Thanks to their digital and above all adaptable corporate architecture, these platforms are not only able to respond to the challenges of the digital age: they even dictate the pace of many online innovations for us consumers. The platforms referred to as "digital ecosystems" are known for their so-called "walled garden" monetarisation strategies.

In short, their recipe for success is: the longer consumers remain on a single platform, the more firmly they become locked in and the more simply the various monetarisation strategies employed on each platform can be translated into lucrative profits. This makes everyday life for us as consumers increasingly easier, as we know that many of our (digital) needs can be met from a single source. Moreover, platform architectures help to transcend conventional hierarchical borders, as well as suboptimal silo structures, in order to strike out new paths interlinking communication, software and hardware solutions. At the same time, a range of new technologies and potential business models are constantly being tweaked on an experimental basis across numerous sectors.

Besides scarcely regulated digital platform providers, fintech start-ups too are throwing their hat into the ring. Fintech start-ups are agile, innovative, and able to move faster than large, more inert corporate groups. Their fully digitalised value networks also allow them to scale their business model optimally. Their recipe for success, too, is based on the harmonious interplay between implemented hardware and software. By achieving optimum integration and using compatible and interoperable technologies, as well as appropriate programming interfaces, fintechs are able to link themselves into the value network of digital ecosystems. Consumers are spoilt for choice by the variety of platforms offered by these start-ups, and are served by new and/or complementary products, firstly in the business-to-consumer segment; but also, above all, in the hotly contested business-to-business segment. The ability to link up with different market players means that many innovation-spawning digital technologies are gradually finding their way into traditional companies, where they are evolving into a comparative competitive advantage for Germany (though not only Germany) as a business location.

Analysis and Use of (Customer) Data as a Basis for Digital Business Models
What the professional handling of data on digital platforms shows above all is the potential of big data under almost fully digital conditions: that is, in an ideal environment. For companies like Google, data and data analyses constitute core activities. To improve the results delivered by a search engine, every search request entered has to be stored, enriched with metadata (such as the IP address), and then evaluated using algorithms devised specifically for this purpose. Such data analysis is made possible by the fact that every interaction between the user and the platform operator takes place via a large range of digital channels. This supplies the platform with all the (personal) data in digital form from the outset.

An IT infrastructure developed specifically for this purpose and consisting of a network of powerful, state-of-the-art computer centres naturally enables the platforms to (a) store these data in a structured format and (b) evaluate them in real time if necessary, so that (c) the customer can be offered personalised services. Furthermore, this infrastructure enables data derived from other sources to be seamlessly integrated into already existing databases. For many companies these ideal, almost completely digitalised conditions are more or less a pipe dream. It follows that they face particular difficulties when it comes to replicating similar data analyses, or implementing algorithm-based solutions in a timely fashion.

Banks possess a tremendous amount of valuable data; and one of the things these data offer is the opportunity to explore new ways of addressing customer needs. In general, the current account is the key interface linking banks and their customers. Banks have access to many valuable customer behaviour patterns (in terms of payments, consumption, propensity to save and invest, risk aversion, travel preferences, etc.). It therefore makes sense for them to apply the same data evaluation strategies as those used by large digital platforms, so that they too can offer their customers convenient, one-stop shopping for as many value-added services relating to their finances as possible. After all, intelligent data analyses are the only means of both maximising customer utility and making internal infrastructures leaner and more efficient in the long term.

Use of Cognitive Technologies as a Means of Gaining Competitive Advantage
In future, cognitive, self-learning systems will be instrumental in supporting decision making in a way which is technologically valuable. They will facilitate, for instance, recognition of valuable correlations in customer promotion campaigns, from which customer groups with similar behavioural patterns and similar preferences can be identified ("cluster analysis"). In the end, customers benefit from being addressed individually within each of the various financial services. Based on knowledge of previous customer habits, new (even as yet unimagined) needs can be met.

Cognitive, self-learning systems can also be put to use internally, such as for regulatory requirements in risk management. Statutory requirements, for example, can be automatically reviewed with regard to their impact and implementation; and the new or amended regulatory requirements may subsequently be deployed in the respective business areas. Likewise, in risk management, audits which must be completed for the fulfilment of regulatory requirements may also be automated. Given the growing degree of regulation in the banking sector, cognitive systems can thus shorten cost-intensive processes in the medium to long term, making them more efficient. Moreover, the use of self-learning systems can guarantee that outcomes are continually improved and become 'smarter' with every interaction. These systems will not completely replace humans, but will offer valuable support in areas of increasing complexity.

A company's progress in adapting its corporate architecture appropriately to the digital age brings with it the challenges of implementing advanced data analysis tools. Companies that succeed early on in digitising their upstream and downstream value-added networks as comprehensively as possible will form the foundations required for future algorithm-based data analyses.

Experts estimate that today only 15% of all globally available data are structured, and about 85% unstructured (TNS Infratest GmbH 2012). The figures are similar for banks. To cope with the steadily growing volumes of data and advanced algorithm-based analytical methods, the first step for banks will be to harmonise all available types of data; that is, to make them machine-readable. The conversion of audio, video and image files into standardised machine-readable data (Heuer 2013) is particularly technically challenging.

At this point it must be remembered that for regulatory reasons established banks are not allowed to cross-correlate personal client data between business divisions in the attempt to gain potential insights from newly acquired data sets. Banks have to observe guidelines, brought in to ensure regulatory compliance, which prohibit the exchange of information between individual business divisions managed under different areas of responsibility. This prevents potential conflicts of interest ("Chinese walls"). Of course, these strict regulatory requirements also apply to the underlying IT systems and (customer) data sets.

Data Protection and Data Security as a Comparative Advantage of Traditional Banks

Since many digital transactions, as well as data access, have migrated from desktop PCs to the cloud, and as mobile devices, too, are becoming an increasingly preferred means of data access, IT security is gaining overriding importance in all spheres of life. The release of the Snowden documents in June 2013 acted as an additional driver of uncertainty and the feeling of "no longer being alone" online. Increasing IT security is now an important step for banks, because—especially when it comes to sensitive financial data—customers are (rightly) concerned by data security breaches. This means that data protection and security could be the trump card for traditional banks in the future. For the financial sector must now take advantage, on the one hand, of this imbalance in the development of valuable state-of-the-art internet services, and, on the other, of the vulnerabilities that have come to light concerning the security of data and IT systems. The worrying attitude towards data protection standards reflected in some of the data practices of major platform operators is what makes this opportunity is such a hot potato for banks. In future it will be particularly important also to win over those who do not or who refuse to use digital banking services.

The Herculean Challenge: Migrating Towards a Digital Platform

The reform measures that must now be implemented are presenting traditional banks with perhaps their biggest challenge to date: to remain competitive, banks need to convert their business model into a platform or a digital banking ecosystem. Against the backdrop of emerging digital ecosystems, the financial industry would be well advised not only to keep an eye on the big internet firms, but also to investigate whether those proven strategies can also be implemented in their own business environment.

The issue here is for banks to create a platform as the basis for a digital ecosystem of their own. It is highly likely that many digital ecosystems will continue to expand their collaboration with credit card providers, telecommunications companies, fintech start-ups, and niche providers, entering into strategic alliances in order to capture further market shares in standardised financial services. To minimise the impact of potential cut-throat competition on financial institutions, traditional banks will need to develop a digital ecosystem with their own digital corporate services, and in addition remain open to integrating into existing alliances, or alternatively take steps to form their own alliances.

Traditional Banks Going Solo: An Unlikely Prospect

Whether banks succeed in building a digital platform on their own remains to be seen. But considering the information edge enjoyed by some digital ecosystems, the complex demands on modern algorithm-based banking, and the increasing costs and margin pressure generated by changes in the competitive environment, strategic alliances are the more likely scenario in the medium term. Future competitiveness will depend on how quickly and flexibly traditional banks are able to respond to the challenges of technological progress and the innovations of digital structural change. Seamless implementation of digital processes and structures could also enable banks to boost their enterprise value, as this digital approach could make it easier and cheaper to realise (even temporary) strategic alliances. On top of this, it would help banks to lock more long-term customers into their own platforms.

2 What Does a Digital Banking Platform Look Like?

Modern online banking is considerably more personalised, simple, intuitive and convenient for the customer. Customers themselves, with their secure online accounts, are placed at the heart of the digital banking ecosystem. Digital customer accounts provide users with instant access to a large variety of services, both from the customer's own bank as well as from external providers linked with the banking ecosystem via programming interfaces. A purpose-built banking app store provides customers with access to an array of internal and external financial products and services (see below Fig. 2).

- Behind the screens, banks are linked both with one another and a range of external financial service providers (such as fintechs, insurance companies and retailers), via programming interfaces at the technology level and contracts at the business level. With the help of the banking app store customers can decide—independently, quickly and conveniently—which products and services they would like to access. The different banking apps or web-based financial services can be offered either for a fee or free of charge. In-house algorithms generate customer-targeted recommendations showing ratings and reviews (à la Amazon), making it easier for customers to choose the services they want. Of equal importance to this interactivity is that customers feel they are permanently in a secure IT environment, and able to communicate and act without being watched.
- Essentially, customers desire a discreet but individually configurable and intelligent (i.e. self-learning) financial assistant in the form of an app or other type of web-based access path to their bank; many also wish for a voice activation feature. The idea of the financial assistant is to support customers in all their daily financial business using data and algorithm-based services.
- Here it is essential that customers have access to their bank and/or online banking on all channels, without fail. Comparable to social network platforms, modern online banking will offer a service enabling customers to customise the

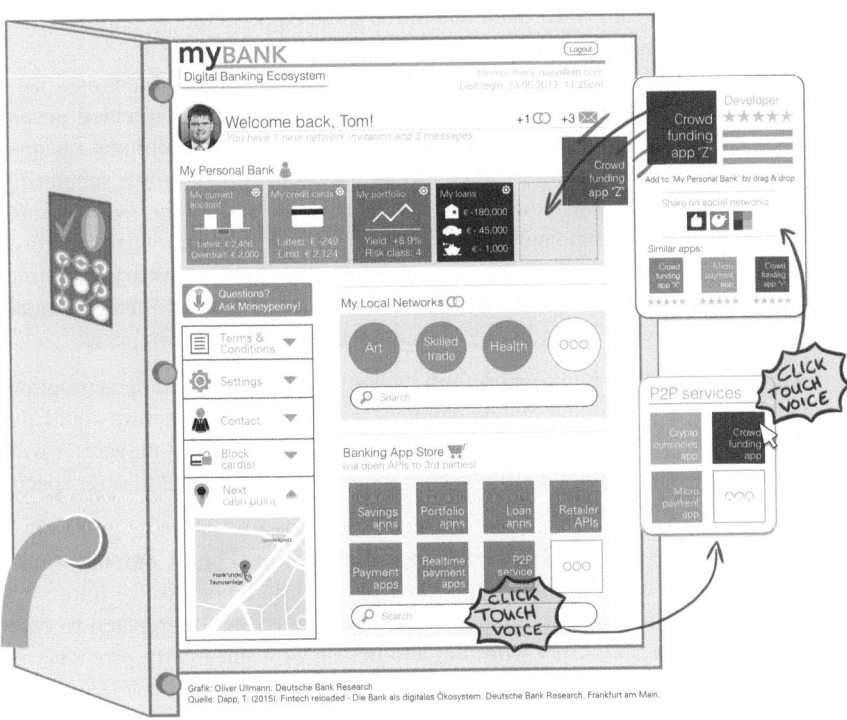

Fig. 2 Digital financial ecosystem (own illustration)

appearance of their own interface, including the ability to freely select frequently used services to display in the foreground—their personal and secure area.

- If customers are on a secure online banking platform and want to configure products and financial services online, bank advisors must be in a position to take continue this configuration—seamlessly—on other channels, without having to reboot systems or re-enter master data. Customers of a modern digital bank should no longer notice that they used different channels in the process of signing a contract or concluding a sale.

- For contract signing in particular, in future banks should also offer exclusively internet-based solutions. As regards authentication on the internet, in future biometric recognition procedures (such as fingerprint, hand vein scans, speech and touch identification procedures) are set to become the norm, and to supplement—if not replace—current identification procedures based exclusively on knowledge and possession.

- Given customer consent and due consideration of the regulatory framework (e.g. banking secrecy principles), differing networks with local links may also emerge, or be actively offered, within the customer's own bank. Banks, for instance, could form a range of interactive networks with local and regional tradespeople or doctors, which could then offer customers their products and services via the banking ecosystem. This would also make for smooth, rapid payment transactions between tradespeople/doctors and customers, since both parties would be linked as customers of the same bank, in theory requiring only one internal transaction entry.

- Another attractive network model is a network which functions as a crowdfunding platform. Some financing projects can be implemented despite being rejected by committees of funding establishments or traditional financial institutions because the "crowd" considers the project to be worth supporting, and provides funding. Creditors and debtors would be customers of the same banking ecosystem, managing their transactions on a bank-owned platform. Here the bank acts merely as the network by providing the necessary infrastructure, and is not liable for potential risks, since the lending of the "crowd capital" is not handled by the bank, but by customers themselves (peer to peer).

As mentioned above, by using uniform technology standards and open programming interfaces, an individual mobile payment service of the banks' own could also be established by joining up other banks, retailers and other market players. A wide array of retailers could link themselves into the banking ecosystem to offer special customer loyalty programmes.

The Fintech Scene Is More Interested in Collaboration Than Confrontation
Digital ecosystems are colliding with each other more and more. In the future it is very possible that market participants will become increasingly prepared to enter into further strategic alliances with one another or with third-party providers by

means of suitable programming interfaces within the value network. The collaboration of Apple (Pay) with several different credit card providers—possibly the best-known technological venture in recent times—is proof of the development of an increasing number of alliances. Among future strategic alliances made, for instance, in the sphere of digital payment services and mobile financial services, international card and payment providers will certainly be just as present as established telecommunications companies. In the past there have, however, been numerous instances of digital ecosystems and Fintech start-ups entering into successful alliances with traditional banks. This is not least because market players from outside the banking sector are in a permanent position to fulfil the strict regulatory requirements which apply within the financial sector, without having to provide their own capacities to do so.

As a rule, these collaborations create synergies and overlaps in terms of size, reach, customers and opportunities for integration and internationalisation. This provides the established banks with opportunities to collaborate with other banks, and also with large internet platforms, small niche operators, or the much-discussed fintech start-ups. They are all operating in the market for digital, data-based and algorithm-based banking. It is also conceivable that there will be strategic partnerships which are able to make complementary offers to expand the range of digital and mobile financial services. Companies across the entire retail sector, as well as certain mobility providers, represent suitable potential collaborators for a digital mobile payment system or for a range of customer loyalty programmes. This would add a further aspect to the banking ecosystem.

The flexible corporate architecture of a digital banking ecosystem enables all expertise available on the financial market to be united at once. This means that modern data and algorithm-based financial services and products can be offered to consumers from a single source, in line with the needs of the internet-savvy customer. Diverse products and services offered by various market players are digitally interlinked and offer customers maximum flexibility in shaping their financial needs. Consumers no longer need to leave the platform; but instead have access to all manner of applications and financial content in the form of apps or web-based services that are tailored to their own personal hardware and software. Moreover, the platform architecture helps to transcend conventional hierarchical boundaries and decades of suboptimal silo thinking within traditional banks, to strike out new paths interlinking communication, software and hardware solutions.

In this way the construction of digital networks no longer gives rise to innovation in purely individual, isolated spheres and sectors; but increasingly also in the open (programming) interfaces involved. In the future, whether an individual player has the necessary expertise and wealth of experience in protected markets will cease to be relevant. The important factor will be the smart links between the diverse infrastructures, skills and abilities of various market participants.

3   Conclusion: Adapt and Optimise "Walled Garden" Monetarisation Strategies in Digital Ecosystems

With their digital infrastructure and their implementation of harmonised, interlinked hardware and software solutions, digital ecosystems are successful market operators. For this reason, the future will bring the most opportunities to those firms and/or banks which succeed in embedding their internal and external processes, products and services early on and as flexibly as possible into a digital company infrastructure, putting them in the position to anticipate swiftly new technologies which could be employed as a platform, or to enter readily into uncomplicated strategic alliances with relevant market players.

In this context, the key to success is a platform policy with suitable program interfaces. This will help to guarantee a flexible corporate architecture in the long run, and allow financial service providers in the future to respond to technological achievements they had not yet imagined. In the future, too, the key to enriching the value-added structure of entities with modern technologies will likely lie in the development or rewriting of software and/or the programming of additional open interfaces (API economy).

The regulatory framework poses a legislative challenge. As already mentioned, banks are confronted with "Chinese walls" which limit the scope of data processing. New competitors, by contrast—and particularly those from outside the banking sector—do not face this problem; and so, in information terms, digital ecosystems continue to enjoy a leading edge in this respect. The result is that, for regulatory reasons, traditional banks permanently lag a step behind in the catch-up process; and so what is required here is a regulatory environment which provides fair rules and a level playing field. This is the only way to guarantee that individual market players are not given preferential treatment to the detriment of traditional banks.

What is more, if they comply ex ante with data protection rules (themselves formulated to satisfy strict regulatory requirements), traditional banks could assume a leading role. Additional self-imposed, i.e. voluntary, measures (e.g. disclosure of the operating methods employed by underlying algorithms) could enable banks to make their analytical practices even more transparent: in contrast to many internet platforms. These confidence-building measures would enable customers to have informed and self-determined knowledge of what happens when their (personal) data is passed on and/or when consenting to an analysis that facilitates their decisions between different financial services. This can also help to overcome the "black box" character of big data.

As long as traditional banks guarantee that they will neither monetarise personal data by selling them to third parties nor misuse them for other non-business projects, they should in future be allowed—with the customer's consent—to conduct data analyses across divisional lines using the information on record. Discussing the issue with customers in advance and documenting their consent will help to create the transparency necessary not only for securing customer trust, but also for complying with data protection legislation with regard to individuals' right to determine the use of their personal information.

Traditional banks now have the opportunity to address themselves to the challenges of structural change in the digital sphere. This does not mean simply adopting a defensive response to change; but rather ensuring that they are perceived as serious, innovative market players, eager to take an active role in the remodelling of financial services. At this juncture, choosing to make the transformation into a banking ecosystem is an effective alternative strategy.

References

Bahr F et al (2012) Schönes neues Internet? Chancen und Risiken für Innovation in digitalen Ökosystemen. Policy Brief 05/12. Stiftung neue Verantwortung. Berlin

Brynjolfsson E, McAfee A (2014) Second machine age. Work, progress, and prosperity in a time of brilliant technologies. W. W. Norton & Company, New York, London

Dapp T (2015) Fintech reloaded—traditional banks as digital ecosystems. With proven walled garden strategies into the future. Current Issues. Digital economy and structural change. Deutsche Bank Research, Frankfurt am Main

Heuer S (2013) Kleine Daten, große Wirkung. Big Data einfach auf den Punkt gebracht. Digitalkompakt LfM #06. Landesanstalt für Medien Nordrhein-Westfalen, Düsseldorf

TNS Infratest GmbH (2012) Quo Vadis Big Data—Herausforderungen—Erfahrungen—Lösungsansätze. TNS Infratest GmbH—Geschäftsbereich Technology, München

A Gift for a Stranger: Freecycling as a Current Lifestyle of Sustainable Consumption

Katharina Klug

1 Introduction

For free! Free software, free songs, free things. The internet offers an excellent platform to share technologies and goods. While **sharing** intangible goods (e.g. sharing music via napster) does not request to give up property at all, tangible products need actually be passed physically (Giesler 2006). Essentially, Freecyclers do exactly this. Freecyclers pass tangible goods and—contrary to existing market mechanism—they do not request any financial compensation. Hence, Freecycling is a modern gift-system (Arsel and Dobscha 2011).

At the time, more and more consumers join the Freecycling initiative to foster sustainable consumption and to life **sustainable** at all. The raising commercialization might be one remarkable reason for the increasing popularity of Freecycling. The community works without using money or trading at all. In contrast, it is based on a pure transfer of tangible good; a transfer between the *good-giving* person and the *good-receiving* person (Grant 2013). Instead of selling needless goods in an extensive way for little cash at local or virtual selling points (e.g. garage sale or ebay), using the Freecycling community means making somebody else happy. Additionally, Freecycling focuses on reducing recourses and surviving with little budget. Those two trends might have the power to become widespread among a whole generation (Lanchester 2015). Therefore, it is usefull to have a look into the Freecycling concept as unconventional form of consumption. Moreover, manager should learn about the consumers' motivation to participate to have a deeper understanding of this upcoming lifestyle.

K. Klug (✉)
Department of Design (AMD), Hochschule Fresenius (University of Applied Science), Munich, Germany
e-mail: katharina.klug@amdnet.de

© Springer International Publishing AG 2017
T. Osburg, C. Lohrmann (eds.), *Sustainability in a Digital World*, CSR, Sustainability, Ethics & Governance, DOI 10.1007/978-3-319-54603-2_17

2 Freecycling: Voluntary-Based Free Recycling

Freecycling is a form of collaborative consumption (short: co-consumption) closely related to the sharing economy. Co-consumption represents a social movement focusing on exchanging, renting, hiring and giving instead of buying tangible (e.g. cars) or intangible goods (e.g. music). Those objectives usually are reached by using digital media to catch a wide audience and to perform in an efficient way (Belk 2014). The Freecycling community was created by Deron Beal in 2003. Meanwhile the uncommercial project www.freecycling.org has become an enormous **gift-network**. Starting with 30 members in Tuscus (Arizona, USA), presently the social movement counts over nine billion Freecyclers acting in about 5,300 regional groups worldwide (76 groups in Germany, 12/2016).

Freecycling stands for nonpaid transfer of consumer goods such as furniture, toys and books (Arsel and Dobscha 2011). The members of the Freecycling community (Freecyclers) may act as both giver and receiver of goods. Freecycler act for selfish as well as for altruistic **reasons** that might be systemized in four categories (Nelson et al. 2007): on the selfish level there is *to declutter* and *to save money* and on the altruistic level there is *to help others* and *to protect the environment* (see Fig. 1). While the giving person benefits from the transfer by retrieving (free) space due to declutter usable (no longer required) items and to become happy due to helping others, the receiver benefits by obtaining useful items for free with a minimum environmental impact. In other words, Freecycling might be considered as a win/win opportunity for all participants with an additional social impact (Hutter et al. 2016). Moreover, the direct communication from gift-giver to

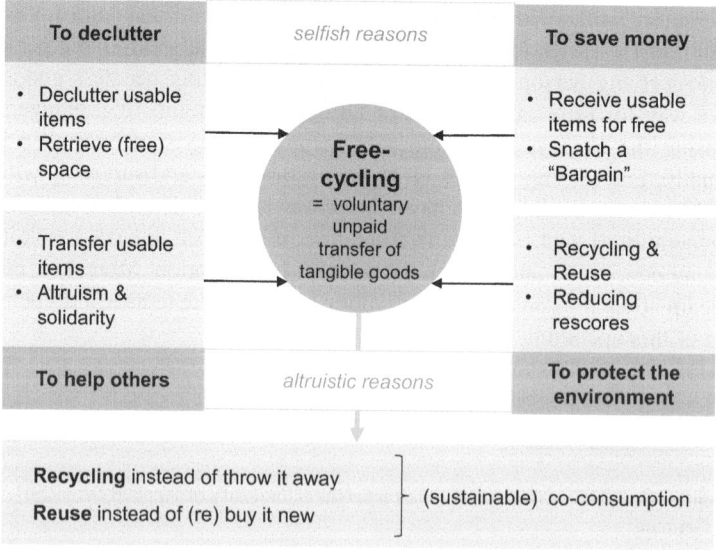

Fig. 1 Underlying motivation to Freecycle

gift-receiver (e.g. while handing over the good) often evokes a mind shift at the giver's perspective to participate in Freecycling away from selfish reasons (e.g. to declutter) towards altruistic motives (e.g. to make someone happy). Frequently, this mind shift even results in a deeper connection with the Freecycling community (Aptekar 2016). In a nutshell, Freecycling connects two central aspects: First, recycling goods instead of throwing it away on the giver side. Second, reuse goods instead of (re-)buying it on the receiver side. Moreover, Freecycling as an unconventional form of co-consumption fosters a consequent sustainability by its regional character (e.g. short transport distances) (Eden 2015).

To keep the system in balance it is necessary that there are enough gift-giver and that gift-receiver do not exploit the system. Due to various participants' motives and types of people creating an **identity** among the community members is the key to a working Freecycling concept. On the one hand, the viability of the Freecycling system is visible in its continuous increasing number of members over 13 years since the initiative was founded. On the other hand, scientific studies confirm a deep-rooted identification and a strong solidarity among the Freecyclers (Grant 2013; Willer et al. 2012). Accordingly, a giving person gives about 21 items to the network. A receiving person (with no actual duty to return anything) gives about nine things to the Freecycling community anyway. Especially, the group of young parents (e.g. benefitting from Freecycling gifts for financially reason) feel a "voluntary" duty to pass received (no longer required) goods to other parents in similar situations. Passing those goods does not mean to lose something valuable. Rather they are fulfilling the community's (invisible) identity norm to (morally) be able (in future time) receiving further gifts from the community.

Freecycling works according to the psychology of gift-giving using the principle of **reciprocity** for human action (Adloff and Mau 2005; Giesler 2006). The literature usually differs between various forms of reciprocity (see Fig. 2), reaching from *negative* reciprocity (= the giving person expects an immediate and direct return from the receiving person) over *balanced* reciprocity (= the giving person expects some return from the receiving person in future time) to *generalized* reciprocity (= the giving person does not expect any return from the receiving person). In that way reciprocity defines the moral standards of human coexistence and determines the flow of giving, receiving and giving (back) (Nelson and Rademacher 2009; Giesler 2006). Due to its consequent non-profit orientation the Freecycling concept is classified to generalized reciprocity, i.e. reasons aside profit and commercialization are more relevant for Freecyclers. Gifts given in mind of generalizes reciprocity (without any expectations) enables the giving person to slip into the receiver's perspective and to act rather altruistic. In this way, in contrast to ordinary sharing platform (such as airbnb or uber), there is no direct exchange relationship between the giving Freecycler and the receiving Freecycler (Grant 2013). Rather, there is a relationship between a single Freecycler and the Freecycling community. Ideally, the transfer of goods processes as follows: person A passes to person B, person B passes to person C etc. A Freecycler receiving various goods from different persons do not attribute the benefit to the single giving person but to the group of Freecyclers. Hence, the community is perceived as

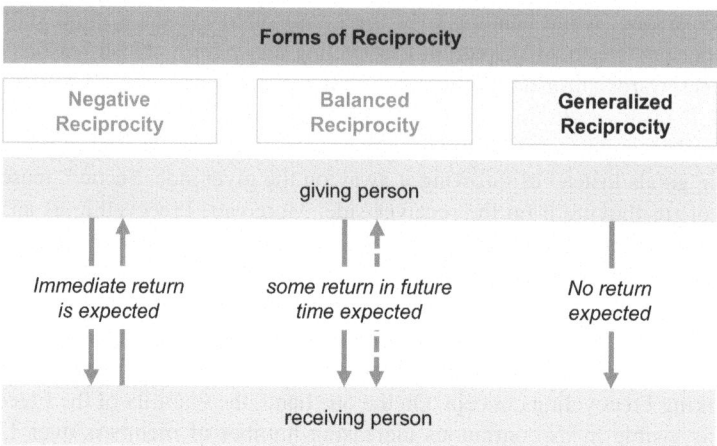

Fig. 2 Forms of reciprocity

source of the gift and gets the receiver's gratefulness. This underlying mechanism step by step ensures merging strangers to a strong community that builds an own (growing) identity. Usually, strengthening the community's identity goes along with more generous gifts provided by the members to the network (Willer et al. 2012).

3   Implications of Co-consumption and Freecycling for Marketers

Sharing is one central characteristic of collaborative consumption respectively the sharing economy. This central idea nowadays is already part of highly efficient business models across all branches of industry. Next to taxi apps (e.g. uber), there are car sharings (e.g, DriveNow by BMW), homestay-networks (e.g. airbnb), travelers' communities (e.g. couchsurfing) and rental platforms (e.g. LeihDirWas) successfully using the digital sharing approach. Those companies' intention is to settle **using** a product instead of buying it. According to current statistics, this sharing approach is interesting and becomes omnipresent especially for the young generation. People up to 30 years old are particularly open minded towards co-consumption intending to intensify its use (Statista 2014). Even if the sharing economy has been criticized for its growing single focus on returns on investments (e.g. Slee 2016), it is still an interesting approach for marketers. Considering Freecycling as an unconventional form of co-consuming there are two learnings marketers should consider:

- **Common use of products:** According to a permanent and more and more rapid changing process in both, private and business environment, the planning horizon for consumer decisions is becoming much shorter than years ago (Matzler et al. 2016). Hence, long term binding decisions are less and less attractive for consumers, unless there is enough flexibility for transferable usage or short term sharing (e.g. subscription for a printed newspaper is transferable to a neighbor during vacation absence).
- **Interaction between consumers:** Private individuals more and more become (autonomous) small businesses, e.g. by renting out rooms or offering a lift. Visibly, an increasing number of people make own (autonomous) decisions mirroring an incremental stronger consumer power (Hoffmann and Hutter 2012). Due to this development, companies are forced to more actively embed consumers into business processes.

While the sharing economy uses the sharing approach for monetary reasons, the Freecycling concept follows non-monetary objectives. Nevertheless, Freecycling with its core assumption *"One Person's Trash Is Another's Treasure"* is implementable also for companies in a specific manner. For example, by establishing a so-called **corporate givebox**. This box allows the companies' staff giving goods to colleagues in a voluntary and anonymous way. Assumedly, a simple shelf placed in an accessible central location within the company plus a short description of the givebox concept is enough to start (see Fig. 3).

While the gift-giver is happy about the new space and his good deed for a colleague, the gift-receiver is pleased about a useful item for free and simultaneously supporting environmental issues. Furthermore, the company offers a platform for common use of products and interactive exchange among personnel focusing on solidarity. As explained above, one the one hand the givebox might contribute to a higher satisfaction level of the individual employee acting as a Freecycler. One the other hand the givebox might foster the company's identity.

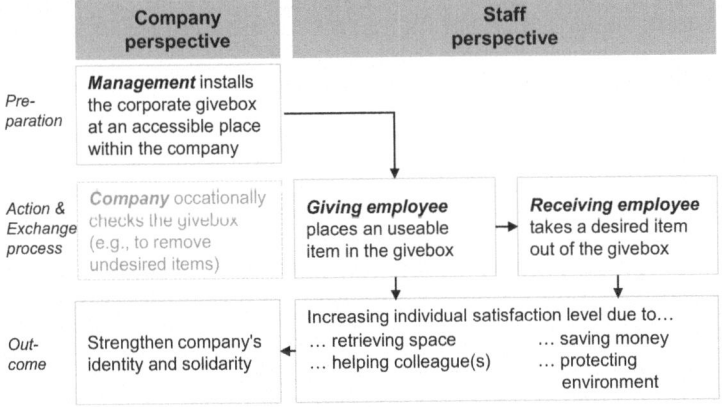

Fig. 3 Process and effects of a corporate givebox

4   Conclusion

It has been shown that Freecycling is a consequent way of sustainable consumption using an online-based community. From a technical perspective the Freecycling community is a simple local mailing list. From an ecological perspective Freecycling has got the power to widespread the recycling and reuse approach. From a social perspective Freecycling even has an identity effect and creates solidarity among the community members.

The pure profit-orientation combined with aggressive marketing strategies made many consumers believe that the primary objective of companies is to raise company profits at the costs of consumers and society (Petrus and Adamek 1988). From the consumers' perspective, Freecycling might be considered as one possible way out; as social movement with a positive core focusing on social well-being (Hutter et al. 2016). From the companies' perspective the forceful non-commercial approach of Freecycling on the first glace might be not interesting for business. Indeed, the opposite is right. As the strict absence of monetary and trading aspects is the central motivation for consumer to participate in Freecycling. For them it represents the authentic roots of co-consumption, that Botsman and Rogers (2010) named humanity and that the current sharing economy sometimes loses sight of the face. Therefore, companies might use the Freecycling approach to foster identity and solidarity e.g. by just installing a corporate givebox.

References

Adloff F, Mau S (2005) Vom Geben und Nehmen. Zur Soziologie der Reziprozität. Campus, Frankfurt am Main

Aptekar S (2016) Gifts among strangers: the social organization of freecycle giving social problems. Soc Probl 63(2):266–283

Arsel Z, Dobscha S (2011) Hybrid pro-social exchange systems: the case of freecycle. In: Ahluwalia R, Chartrand TL, Ratner RK (eds) NA—advances in consumer research, vol 39. Association for Consumer Research, Duluth, MN, pp 66–67

Belk R (2014) You are what you can access: sharing and collaborative consumption online. J Bus Res 67(8):1595–1600

Botsman R, Rogers R (2010) What's mine is yours: the rise of collaborative consumption. Harper Collins, London

Eden S (2015) Blurring the boundaries: prosumption, circularity and online sustainable consumption through freecycle. J Consum Cult, online first. doi:10.1177/1469540515586871

Giesler M (2006) Consumer gift systems. J Consum Res 33(2):283–290

Grant A (2013) Geben und Nehme: erfolgreich sein zum Vorteil aller. Droemer, München

Hoffmann S, Hutter K (2012) Carrotmob as a new form of ethical consumption. The nature of the concept and avenues for future research. J Consum Pol 35(2):215–236

Hutter K, Mai R, Hoffmann S (2016) Carrotmob: a win–win–win approach to creating benefits for consumers, business, and society at large. J Bus Soc 55(7):1059–1077

Lanchester J (2015) Die Sprache des Geldes: und warum wir sie nicht verstehen (sollen). Klett, Stuttgart

Matzler K, Veider V, Kathan W (2016) Collaborative consumption: Teilen statt Besitzen, Wie Unternehmen das Phänomen der Sharing Economy für sich nutzen können. In: Granig P, Hartlieb E, Lingenhel D (eds) Geschäftsmodellinnovationen—Vom Trend zum Geschäftsmodell. Springer Gabler, Wiesbaden

Nelson MR, Rademacher MA (2009) From trash to treasure: freecycle.org as a case of generalized reciprocity. Adv Consum Res 36:905–906

Nelson MR, Rademacher MA, Paek HJ (2007) Downshifting consumer = upshifting citizen? An examination of a local freecycle community. Ann Am Acad Pol Soc Sci 611(1):141–156

Petrus G, Adamek RJ (1988) Taking the role of the other: an aid to marketing applied sociology. Teach Sociol 16(1):25–33

Slee T (2016) Deins ist Meins, Die unbequemen Wahrheiten der Sharing Economy. Kunstmann, München

Statista (2014) Nutzung des Sharing Economy Angebote. https://de.statista.com/statistik/daten/studie/330247/umfrage/welche-angebote-aus-der-share-economy-die-deutschen-nutzen/, 08.12.2016

Willer R, Flynn FJ, Zak S (2012) Structure, identity and solidarity: a comparative field study of direct and generalized exchange. Adm Sci Q 57:119–155

Part III
Participation, Education and CSR

Online Learning—Do MOOCs Contribute to the Goals of Agenda 21: "Education for Sustainable Development"?

Christiane Lohrmann

1 Introduction

Digital innovations are rapidly changing learning and education. Academic institutions and the corporate learning market have especially been affected. This article provides insight into the education market, which is rapidly transforming due to challenges presented by digital change, demographic development and new communication habits in the context of Web 2.0. Special focus will be placed on the role of academic online platforms offering Massive Open Online Courses (MOOCs), as they provide open access to learning experiences in order to further the education and careers of anyone with an internet connection. Does this development have the potential to offer substantial education to people worldwide? And can MOOCs eventually contribute to a more sustainable educational development?

This article begins by examining the development of the political concept of *sustainability* and describing how the idea of education for sustainable development (ESD) evolved into the Rio Declaration of 1992. In this context, the article looks particularly at the notion of access to education as a premise for acquiring knowledge for sustainable development and for identifying the problems of non-sustainable development.

This article then takes a deep look into how the educational market has changed as well as how it has dealt with the implications of digital change and demographic development. It explains that, in recent years, millions of people worldwide have discovered the internet as a means to access higher education and use MOOCs to further their education and careers. Academic knowledge formerly only accessible to a small group of people is today free and open to everyone and, therefore, offers new possibilities to further education worldwide. At the same time, research has

C. Lohrmann (✉)
FranklinCovey Leadership Institut, Munich, Germany
e-mail: post@christianelohrmann.de

revealed that only a small percentage of MOOC students are completing their courses. Therefore, questions remain: What kind of impact do MOOCs *really* have and can they contribute at all to the ambitious goals of ESD? Are MOOCs merely an intellectual diversion for the well-educated and well-off? What kind of impact do MOOCs have in both developed and undeveloped countries? Lastly, who benefits from taking a MOOC? This article analyses the correlation between the new phenomenon of *MOOCs* and the impact they have on their learners' lives and careers in the context of ESD. Can MOOCs contribute to the concept of ESD?

2 Sustainability and Sustainable Development: Evolution of a Concept

The concepts of *sustainability* and *sustainable development* have evolved since the 1970s into political and social keywords that have a normative influence on individual and social activities. Sustainability as an issue was first formulated in the eighteenth century in the context of forestry and the mining industry in Saxony by politician and chamber member Hans von Carlowitz. Today, it stands for social, economic and ecological responsibility in society (Carlowitz 2013; RNE 2012). In 1972, the Club of Rome picked up on the subject and drew attention to the 'Limits to Growth'. This was followed in 1987 by the UN World Commission on Environment and Development's (Brundtland Commission) report 'Our Common Future', which for the first time offered a generally recognised definition of sustainability: 'Humanity has the ability to make developments sustainable—to ensure that it meets the needs of the present without compromising the ability of future generations to meet their own'. (WCED 1987, 9). First defined in the Rio Declaration of 1992, the paradigm of sustainability has been sharpened in subsequent UN conferences. Today it is an element of most national constitutions, international treaties and laws. The paradigm of sustainability has become as important as the principles of democracy and the rule of law and is one 'of the key questions for rule of law legitimacy at the beginning of the twenty-first century'.

With the Brundtland Report of 1987 and Agenda 21, the final document of the Conference of Environment and Development in Rio de Janeiro in 1992, the notion of sustainable development was gradually introduced into the international dialogue by the so-called Brundtland Report: 'Sustainable development is (...) a process of change in which exploitation of resources, the direction of investments, the orientation of technological development, and institutional change are made consistent with future as well as present needs' (WCED 1987, 9).

Sustainable development is not a descriptive model but a normative concept. It conveys a picture of the world as it should be, especially one in which more intergenerational and intragenerational justice in terms of the distribution of goods and services is provided (Enquete-Commission of German Bundestag des Deutschen Bundestages 12.6.2002).

Fig. 1 The three
dimensions of sustainability

Provided that we are jointly working for a sustainable world and society, we will have to follow a process that strives towards sustainable development. Here, economic, ecological and social goals must be considered equal (Schaltegger et al. 2003, 185). Sustainability is therefore multidimensional and can be reached by integrated development towards social, ecological and economic goals. In this context, ESD becomes important, as it has a significant impact on all three dimensions of the triple bottom line (see Fig. 1).

3 Education for Sustainable Development: The Rio Concept

The 1992 Rio Conference document acknowledges, for the first time and on an international level, the importance of education for sustainable development: 'Education, raising of public awareness and training, is linked to virtually all areas in Agenda 21, and even more closely to the ones on meeting basic needs, capacity building, data and information, science, and the role of major groups' (Agenda 21, Article 36, 1992).

The most important message conveyed by Article 36 is that the UN widened the concept of sustainable education from ecological education to the broader concept of education for sustainable development. It claimed that in order to support the process of sustainable development, a change of thinking and paradigm was needed, which would require an education strategy that has in mind to qualify societies to contribute to sustainable development (Bormann and de Haan 2008). This includes an awareness of questions in the economic, social, technical and cultural fields. As a whole, ESD puts forward a modern and broad concept of education that is focused on the issue of creating a sustainable future. Here, governments are encouraged to consult with people in isolated situations, whether geographically, culturally or socially, to ascertain their needs and train them to contribute more fully to developing sustainable work practices and lifestyles (Agenda 21, 36.20). Therefore, one central premise for the development of ESD is access to education and, consequently, the potential to participate in training and learning. The concept of ESD has essential characteristics that can be implemented in many culturally appropriate forms; for instance, ESD:

- is based on the principles and values that underlie sustainable development;
- deals with the well-being of all four dimensions of sustainability—environment, society, culture and economy;
- uses a variety of pedagogical techniques that promote participatory learning and higher-order thinking skills;
- promotes lifelong learning;
- is locally relevant and culturally appropriate;
- is based on local needs, perceptions and conditions, but acknowledges that fulfilling local needs often has international effects and consequences;
- engages formal, non-formal and informal education;
- accommodates the evolving nature of the concept of sustainability;
- addresses content, taking into account context, global issues and local priorities;
- builds the civil capacity for community-based decision making, social tolerance, environmental stewardship, an adaptable workforce, and a good quality of life;
- is interdisciplinary. No single discipline can claim ESD for itself, and all disciplines can contribute to ESD (UNESCO 2004).

The concept of ESD is highly ambitious and combined with many hopes for sustainable development worldwide. At the world summit in Johannesburg in 2002, it was again pointed out how important education is for sustainable development in the contemporary world. As recommended by the world summit, the UN General Assembly declared 2005–2014 the Decade of Education for Sustainable Development (UNDESD). Coordinated by UNESCO, the UNDESD emphasised education as a premise for individual and social development in the context of sustainability (Mayer-Schönberger and Cukier 2014, 11).

The progress towards sustainable development basically relies on self-activity and the capability of reflection. It focuses on new forms of self-organised and project-orientated learning that help people understand complex contexts and see themselves as part of a society capable of dealing with problems in a critical, productive and constructive way. In this context, the ability of *Gestaltungskompetenz* (structuring, forming and shaping) becomes relevant (de Haan 1999). De Haan defined *Gestaltungskompetenz* as 'the ability to shape the future of society in which one lives by actively contributing to the notion of sustainable development' (de Haan and Harenberg 1999, 16). This implies the following 11 partial competences (Bormann and de Haan, 2008, 62):

1. Competence of anticipation: thinking and acting ahead
2. Competence of working interdisciplinarily: understanding complex systems
3. Competence of taking the other's perspective: understanding worldwide phenomena in their impacts and relationships

4. Competence of handling incomplete and complex information: ability to deal with risks, dangers and insecurities
5. Competence of participation: ability to participate in processes of sustainable development, structuring, forming and shaping
6. Competence of cooperation: ability to cooperate in planning and acting
7. Competence of dealing with individual decision dilemmas: reflecting on conflicting goals in decision processes
8. Competence of empathy and solidarity: engagement for more justice
9. Competence of self-motivation: shaping of a sustainable future
10. Competence of reflection about individual and cultural role models: critical reflection of one's own actions in relation to social and cultural role models
11. Competence of acting morally: reflecting on justice as a basis for decisions for sustainable development

To promote the process of sustainable development, a change of thinking and awareness is necessary. In this context, a strategy for education towards sustainable development with an emphasis on self-organised and project-orientated learning plays an important role. Knowledge and cognition psychologists stress that the learning context is essential for the ability to act in a responsible way. To have access to and be able to participate in education has a strong impact on everyone's future (Stoltenberg and Rieckmann 2011, 117ff; Rifkin 2015, 163). In this respect, the digital development can create new perspectives by offering access to online education to a broad segment of the global population and therefore potentially overlap with the goals of education for sustainable development. More than one-third of the global population have access to relatively affordable mobile phones and the internet and are therefore able to share information in an increasingly interconnected and collaborative world (Rifkin 2015, 14).

4 Education in the Digital World: The Influence of the Social Web

The internet is revolutionising education and has the potential to outpace classical classroom education to some extent. Similar to the historical development of the printing press, cameras, radio and television, the internet is currently revolutionising traditional structures of communication and learning. Back in the 1930s, German author Berthold Brecht aimed 'to develop the radio from a distribution channel to a communication machine' (Brecht 1967, 127ff). In the 1970s, publicist Hans Magnus Enzensperger picked up on this thought by calling for interactive television (Enzensberger 1997). Later, in the 1990s (the early internet age), the American academic and computer scientist Nicolas Negroponte created the term *digital revolution* (Negroponte 1995). With the development of Web 2.0, traditional structures of communication have again been transformed, especially as O'Reilly put it:

> Software as a continually-updated service gets better the more people use it, consuming and remixing data from multiple sources, including individual users, while providing their own data and services in a form that allows remixing by others, creating network effects through an "architecture of participation" and going beyond the page metaphor of Web 1.0 to deliver rich user experiences. (O'Reilly 2005)

Web 2.0 has enabled internet users to not only consume information but also act as *prosumers* who create content. This, in turn, serves as a possible premise for participation.

5   Access to the Internet: Premise for Participation

Access to education has improved in recent years because of the digitalisation of nearly every area of life. More than one-third of the world's inhabitants have access to relatively affordable mobile phones and PCs and are therefore able to share information though video, audio and text in a collaborative and digitised way. More than six million people worldwide are already registered in MOOCs in order to obtain diplomas from very well-known universities (Rifkin 2015, 14). Does this trend lead to more ESD?

Until a couple of years ago, ESD was primarily examined in an organisational context of creating networks for sustainable development (vgl. Müller-Christ 2014). Today, however, digitalisation has impacted the educational sector at a fundamental level, since the technological progress afforded by Web 2.0 and social media is making radical change possible, especially by meeting the increasing demand for digital learning (Hüther 2014). This development has been particularly pushed forward by demographic factors, including the increasing number of students (see also Metzner 2014 in: Lohrmann 2014a; Kroker 2014 in: Lohrmann 2014c; Koller 2014 in: Lohrmann 2014b). At the same time, the number of people using smart phones, tablets and PCs has also risen exponentially. According to the Roland Berger research study, 'Think Act', the number of people using smart phones worldwide will triple between 2013 and 2019 to more than 5.5 billion people. In the same period, mobile traffic via smart phones is expected to increase tenfold, while the cost for such technologies will decrease significantly (Statista 2014; Reinhold 2014, 5ff).

This development will only improve the potential for online learning via mobile devices. 'Digital natives' are accustomed to communicating over the internet and having 'on demand' access to a wide variety of available content (see Bischoff et al. 2013).

Companies worldwide are watching this development and, according to the 'Think Act' study, spent around US$ 210 billion in 2011 on corporate learning, with 20% of this investment devoted to corporate e-learning (Reinhold 2014, 5). Increasing venture capital investments in e-learning also indicate the positive expectations of online learning. Whereas US$52 million was invested globally in technological innovations in 2005, by 2012 investment had increased twentyfold to

1.1 billion, 60% of which originated in the US (Reinhold 2014, 6). Combined with the development of social media, these investments in e-learning could substantially change the demand for online education.

At the same time, technological innovations have made online education more easily available; through freemium business models, for example. Moreover, transmission bandwidth and speed as well as server capacities and data speed have improved substantially. At both the national and EU level, different education projects are being promoted, such as the 'Open Education Europe' (OER) initiative (vgl. EU Commission—Joint Research Centre 2014). Lastly, the EU has created, via ECTS (European Credit Transfer and Accumulation System), a common currency for education certificates that has helped set common standards across Europe.

6 MOOCs: Change Agents for Education for Sustainable Development?

In 2011, several top universities began experimenting with Massive Open Online Courses (MOOCs). Institutions like Yale University, Harvard University, the University of Michigan, and the University of Pennsylvania opened their digital doors to anyone with an internet connection, promising to deliver unlimited access to high-quality education. A MOOC is a model for delivering learning content online to any person who wants to take a course, with no limit on attendance. MOOCs are learning formats that can be repeated individually and which combine traditional forms of learning through videos and exams with peer coaching. This format has helped open the previously closed system of universities and colleges, and has thus also enabled academics to reach more students and learners than ever before (vgl. Thrun 2014).

Among the most important MOOC platforms are Coursera and Udacity (both based in Stanford University) and edX (Harvard MIT). Each of these platforms evolved from a university environment and have been set up as non-profit organisations and social start-ups. Most of these platforms offer education in a so-called *freemium model*, which means that basic education modules are free but students must pay for further education as well as certificates or diplomas (see Fig. 2).

In the last 3 years, over 25 million people from around the world have enrolled in MOOCs offered by Coursera, edX, and other platforms. Given this impressive number, academics and the media have been surprised by the development: The phenomenon was even made a *New York Times* title story in 2012, 'Year of the MOOCs' (Pappano 2.11.2012). German neuroscientist Gerald Hüther pointed out that MOOCs are opening up new ways of learning that the state system has not offered until now: 'MOOCs live off the motivation of students to learn. . . . Schools are relics of the industrial age' (Hüther in: Lohrmann 2014d). In spite of the impressive number of MOOC students, however, the question remains: Are

Name	Founders	Partners	Courses	Users
Coursera	2012 A. Ng, D. Koller	108 Universities	627	6,400,000
OpenupED	2013 EADTU	11 Universities, EU	272	k. a.
edX	2012 MIT, Harvard	31 Universities	133	1,800,000
Future-Learn	2013 Open University	26 Universities	44	200,000
Udacity	2012 S. Thrun et al.	3 Universities	33	1,800,000

Fig. 2 The largest MOOC platforms (Reinhold 2014, 8)

MOOCs really helping to develop education in the context of *Gestaltungskompetenz* (shaping competence), as called for in the UN's 1992 Rio Conference and its declaration of Education for Sustainable Development?

7   MOOC Students: Who Are They, Where Are They from, and What Do They Learn?

Who are the MOOC students and where do they come from? According to Coursera research, approximately 60% come from developed countries (defined as members of the OECD). As shown below, it is significant that most MOOC learners come from BRIC countries like India and China (see Fig. 3, also Rifkin 2015, 171ff).

However, sceptics are not convinced about the quality of learning output and doubt that the results really contribute to what ESD means (Schulmeister 2013, 7ff). While on the one hand, the number of course learners has increased worldwide; on the other, the number of students who quit mid-course is extremely high: around 90% (Werner 2014). Only 4% of Coursera users who watch at least one course lecture go on to complete the course and receive a credential. Still, given the large number of registered users, the absolute reach of MOOCs is still relevant. With information and communication technologies (ICT) evolving in both developed and bottom of the pyramid (BOP) markets, its economic and social impacts are becoming evident, particularly in BOP regions. By having access to the internet, all users have the opportunity for online education (Rifkin 2015, 7). As German neuroscientist Gerald Hüther put it: 'MOOCs offer new possibilities of learning that the state system doesn't provide' (Hüther 2014 in: Lohrmann 2014d). However, the question remains whether this type of innovative education platform is merely a success in terms of student numbers or whether it actually contributes a valid and high-quality learning experience that will eventually lead to ESD.

Nearly 52,000 people responded to a Coursera survey of 780,000 people from 212 countries who had completed a Coursera MOOC: 58% of respondents were employed full time, while 22% were full-time or part-time students in a traditional

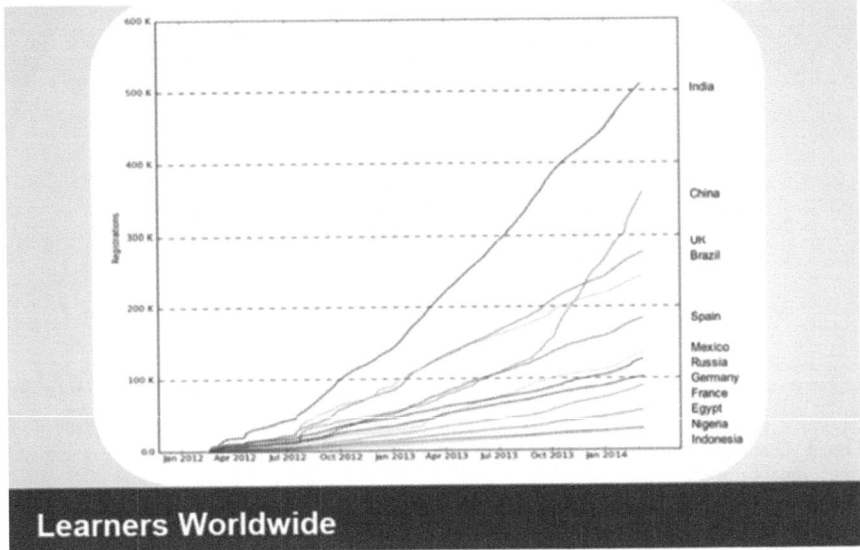

Learners Worldwide

Fig. 3 Development of MOOC students worldwide (Koller 2014)

academic setting. In addition, 83% had at least a bachelor's degree. Thirty-four percent of respondents were from the US, 39% were from other OECD countries, and 26% were from non-OECD countries. The majority of respondents reported career or educational benefits, and a substantial proportion reported tangible benefits such as getting a new job, starting a business, or completing prerequisites for an academic programme (HBR 2015). Twenty-eight percent of those who completed a MOOC enrolled primarily to achieve an academic goal—the so-called 'education seekers'. Below are some of the results the Coursera study (see Fig. 4). Among non-student MOOC completers, those with a lower socioeconomic status and lower levels of education, as well as those from developing countries, were all more likely to report educational benefits, such as received credit or waived prerequisites for an academic programme.

Among the education seekers who are not in a traditional academic setting, disadvantaged populations were more likely to report educational benefits. Furthermore, education seekers from developing countries were more likely to report educational benefits, and those with a low socioeconomic status were more likely to report benefits than those with a higher status. Finally, MOOC students without a postgraduate degree were more likely to report benefits than those who had one. Of these education seekers, 88% reported an educational benefit of some kind. Eighty-seven percent reported an intangible educational benefit (such as gaining knowledge in their field), and 18% reported a tangible educational benefit, such as either gaining credit towards an academic degree or completing prerequisites for an academic programme (see Fig. 5, HBR 2015).

Learners with no postgraduate degree, from low SES brackets, and from emerging economies are more likely to report educational benefits.

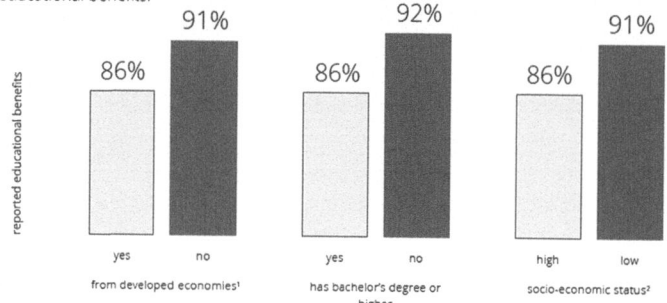

¹Developed and emerging economies are evaluated using indicators from the Organisation for Economic Co-operation and Development (OECD).

²SES, or socioeconomic status, is evaluated as a combination of factors including income, level of education, and occupation. SES was self-reported by respondents.

Fig. 4 Learners regardless of educational, geographic or economic background can expand educational horizons (Coursera 2014)

The Educational Benefits of MOOCs

As reported by those who stated educational benefits as their primary reason for completing a MOOC.

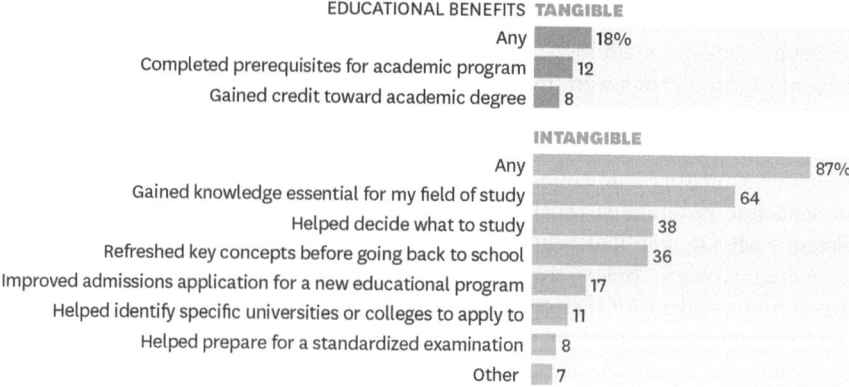

Fig. 5 The educational benefits of MOOCs. Source: Coursera survey data, © HBR.ORG

Furthermore, people from developing countries are more likely to be education seekers, as are people with a lower socioeconomic status. Economically and academically disadvantaged populations are taking particular advantage of MOOCs. In addition, learners from developing countries as well as people with lower levels of education and a lower socioeconomic status also reported tangible career benefits, such as finding a job, starting a business, or receiving a raise or promotion (see Fig. 6, HBR 2015).

Who's Getting Ahead at Work Because of MOOCs?

Tangible career benefits reported by those who stated such benefits as their primary reason for completing a MOOC.

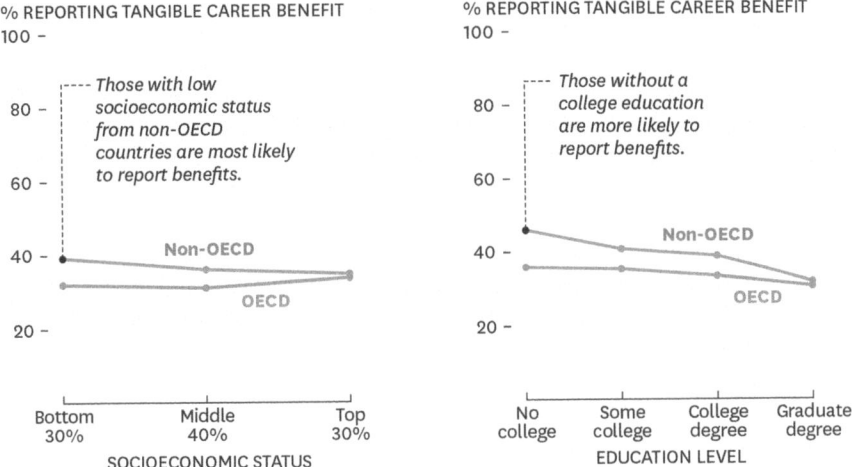

Fig. 6 Who's getting ahead at work because of MOOCs? Source: Coursera survey data, © HBR. ORG

8 Conclusion: Do MOOCs Contribute to the Goals of Agenda 21: 'Education for Sustainable Development'?

The study results of the 'Learner Outcomes in Open Online Courses 2015' survey support the hypothesis that MOOCs can contribute to the goals of ESD as formulated in Agenda 21: 'Countries could support university and other tertiary activities and networks for environmental and development education. Crossdisciplinary courses could be made available to all students and provide an opportunity for those who are less advantaged and have limited access to education' (Agenda 2, 36.5, i).

Of course, MOOCs are available only to those people who have access to the internet, and completion rates remain low. However, over the last few years, more than one million people have completed courses from Coursera alone, and more than 100,000 people have certified completion from Harvardx and MITx courses, from which it can be concluded that many of them derived career or educational benefits from the opportunity (HBR 2015). Moreover, the number of people with internet access or smart phones continues to rise significantly.

Continued innovation and research therefore needs to focus on two main issues: First, how can the quality of learning output be guaranteed? and second, how can students be supported to complete MOOCs so as to secure desired career and educational benefits?

The Coursera survey data illustrate the possibilities MOOCs offer to change the educational landscape and make a substantial contribution to the normative concept of ESD. The courses are reaching large numbers of people, and disadvantaged learners are more likely to report tangible benefits. Also, many of the research findings offer a contribution to the competences claimed to help create *Gestaltungskompetenz* and contribute to ESD, such as the competence of anticipation, of thinking and acting ahead, of gaining knowledge of one's field of study, and of working interdisciplinarily to complete requisites for an academic programme (Zhenghao et al. 2015, in: HBR 2015; Stoltenberg and Rieckmann 2011; de Haan 2008).

MOOCs should be taken seriously as platforms that increase access to education and therefore play a collective and cooperative role. According to sociologist Jeremy Rifkin, they make a substantial contribution towards rethinking the educational process in the Digital Age (Rifkin 2015, 163); a position also supported by Hüther, who believed that MOOCs disrupt the existing system as well (Hüther in: Lohrmann 2014d).

At the same time, MOOCs certainly do not solve all the myriad problems of global education, nor will they replace face-to-face learning as supported by some publicists. However, they have created increased awareness, especially because of the high number of registered students. So, all in all, MOOCs can be considered a step in the right direction, providing open access to learning experiences, especially for those people with limited access to classical face-to-face education, many of whom live in non-OECD countries where MOOCs are seen as beneficial for furthering education and careers.

References

Bischoff L, Friedrich J-D, Müller, Ulrich, Müller-Eiselt R, von Stuckrad T (Hg.) (2013) Die schlafende Revolution. 10 Thesen zur Digitalisierung

Brecht B (1967) Gesammelte Werke. Schriften zu Literatur und Kunst 1. Suhrkamp, Frankfurt am Main

Coursera (2014) Impact revealed—learners outcome in open online courses. https://d396qusza40orc.cloudfront.net/learninghubs/LOS_final%209-21-.pdf. Accessed 10 May 2016

De Haan G (1999) Die Kernthemen der Bildung für nachhaltige Entwicklung, Zeitschrift für internationale Bildungsforschung und Entwicklungspädagogik, 25/Issue 1 "Rio—10 Jahre nach dem Weltgipfel"

De Haan G (2008) Gestaltungskompetenz als Kompetenzkonzept der Bildung für nachhaltige Entwicklung. In: Bormann I, de Haan G (eds) Kompetenzen der Bildung für nachhaltige Entwicklung. Operationalisierung, Messung, Rahmenbedingungen, Befunde. VS Verlag für Sozialwissenschaften, Wiesbaden

De Haan G, Harenberg D (1999) Bildung für eine nachhaltige Entwicklung. Gutachten zum Programm. Bund-Länder-Komm. für Bildungsplanung und Forschungsförderung, Bonn

Enquete Kommission des Deutschen Bundestages (2002) Globalisiserung der Weltwirtschaft – Herausforderungen und Antworten. 14/9200. Herausgegeben von der Enquete Kommission des Deutschen Bundestages. Zugriff. http://dip21.bundestag.de/dip21/btd/14/092/1409200.pdf

Enzensberger HM (1997) Baukasten zu einer Theorie der Medien kritische Diskurse zur Pressefreiheit. Fischer, München

Harvard Business Review (2015) Who is benefitting from MOOC's and why. https://hbr.org/2015/09/whos-benefiting-from-moocs-and-why. Accessed 17 Jan 2017

Koller D (2014) Microsoft Powerpoint—Coursera University—June 2014—TUM

Lohrmann C (2014a) I 6. Interview with Joachim Metzner. 15 May 2014. Telephone, Audio and transcript

Lohrmann C (2014b) I 9. Interview with Daphne Koller. 27 June 2014 in Munich, Audio and transcript

Lohrmann C (2014c) I 10. Interview with Britta Kroker. 06 Aug 2014 in Herrsching am Ammersee, Audio and transcript

Lohrmann C (2014d) I 11. Interview with Gerald Hüther. 24 Oct 2014. Telephone, Protocol

Mayer-Schönberger V, Cukier K (2014) Lernen mit Big Data. Die Zukunft der Bildung. Redline Verlag, München

Müller-Christ G (2014) Hochschulen für eine nachhaltige Entwicklung. Netzwerke fördern, Bewusstsein verbinden. Unter Mitarbeit von Anna Katharina Liebscher. In: Deutsche UNESCO-Kommission e.V (ed) Bonn. https://www.hrk.de/fileadmin/redaktion/A4/20140928_UNESCO_Broschuere2014_web.pdf

Negroponte N (1995) Being digital. Knopf, New York

O'Reilly T (2005) Web 2.0: compact definition. O'Reilly Radar. http://radar.oreilly.com/2005/10/web-20-compact-definition.html. Accessed 7 Mar 2017

Pappano L (2012) The year of the MOOCs. Herausgegeben von. New York Times. http://www.nytimes.com/2012/11/04/education/edlife/massive-open-online-courses-are-multiplying-at-a-rapid-pace.html?pagewanted=all&_r=0. Zugriff 7 Aug 2016

Reinhold T (2014) Think act. Corporate learning. Herausgegeben von Roland Berger Strategy Consultants GmbH (Hrsg.). München

Rifkin J (2015) Die Null-Grenzkosten-Gesellschaft. Das Internet der Dinge, kollaboratives Gemeingut und der Rückzug des Kapitalismus. Campus, Frankfurt

RNE: Zehn Jahre Nachhaltigkeitsstrategie: RNE zieht kritische Bilanz. Pressemitteilung vom 18.04.2012. http://www.nachhaltigkeitsrat.de/de/news-nachhaltigkeit/2012/2012-04-19/zehn-jahre-nachhaltigkeitsstrategie-rne-zieht-durchmischte-zwischenbilanz/?size=yjlnmhbvpimaf&blstr=0. Zugriff 7 Aug 2014

Schaltegger S, Burrit R, Petersen H (eds) (2003) An introduction to corporate environmental management. Striving for sustainability. Greenleaf, Sheffield

Schulmeister R (Hg.) (2013) MOOCs – massive open online courses. Offene Bildung oder Geschäftsmodell? Waxmann

Statista (2014) Anzahl der Smartphone-Nutzer in Deutschland in den Jahen 2009 bis 2014 (in Millionen). http://de.statista.com/statistik/daten/studie/198959/umfrage/anzahl-der-smartphonenutzer-in-deutschland-seit-2010/. Accessed 7 Aug 2014

Stoltenberg U, Rieckmann M (2011) Partizipation als zentrales element von Bildung für eine nachhaltige Entwicklung. In: Heinrichs H, Kuhn K, Newig J (eds) Nachhaltige Gesellschaft. VS Verlag für Sozialwissenschaften, Wiesbaden, pp 117–131

Thrun S (2014) MOOCs can help the economy even if they can't replace college. https://gigaom.com/2014/01/25/sebastian-thrun-MOOCs-can-help-the-economy-even-if-they-dont-replace-college/

UNESCO (2004) Education for sustainable development. http://www.unesco.org/new/en/education/themes/leading-the-international-agenda/education-for-sustainable-development/education-for-sustainable-development/, 16 Oct 2016

UNSD: UNCED (1992) Agenda 21. https://sustainabledevelopment.un.org/content/documents/Agenda21.pdf, 7 Aug 2016

von Carlowitz HC (2013) Sylviculutra oeconomica oder Haußwirthliche Nachricht und Naturmäßige Anweisung zur wilden Baum-Zucht. München

WCED (1987) Our common future. Report of the World Commission of Environment and Development

Werner G (February 2014) MOOC – eine Bildungsrevolution? Interview regarding Massive Open Online Courses with Prof. Dr. Rolf Schulmeister, Dr. Frank Hoffmann. Editor: Universität Hamburg, Newsletter. Hamburg. URL: http://www.uni-hamburg.de/newsletter/februar-2014/mooc-eine-bildungsrevolution-interview-zu-massive-open-online-courses-mit-prof-dr-rolf-schulmeister-und-dr-frank-hoffmann.html. Accessed 11 Nov 2014

Zhenghao C, Alcorn D, Christensen G, Eriksson N, Koller D, Emanuel EJ (2015) Who is benefitting MOOCs and why. Harvard Business Review, September 15

Technology Adoption at the BOP Markets: Insights from Turk Telekom's Focus on Inclusive Business

Dicle Yurdakul, Seda Müftügil Yalçın, and Zeynep Gürhan-Canlı

1 Introduction

In this chapter, we examine Turk Telekom's "Life is Simpler with Internet" initiative, which we evaluate as an inclusive business activity that has contributed to both company's sustainable market development and also to sustainable development goals in the Turkish context. Our analysis is informed by a thorough literature review that was conducted on technology and innovation with respect to inclusive business and also a qualitative study based on an in-depth interview. We interviewed Corporate Social Responsibility Manager Tülin Kara Özgen and Corporate Social Responsibility Senior Specialist Hande Gürsoy, both working in the CSR department of Turk Telekom and who have been involved in the planning and execution of the project from the start of the initiative. Through its unique and innovative structure, Turk Telekom, a leading internet service provider in Turkey, was able to embark a holistic and sustainable effort to overcome the "digital gap" in Turkey while growing its business and creating social impact on the ground.

In the following pages, we will focus on the internal dynamics of this project whose main success factors stemmed greatly from the collaborative structure the initiative had, sustainable relations the company established with its new customers and with the skill development component existing in the project. We begin by providing brief information about inclusive business and development and how information and communication technologies (ICT) contribute to economic and human development in base of the pyramid (BOP) markets. Then, we will reveal

D. Yurdakul
Kemerburgaz University, Istanbul, Turkey
e-mail: dicle.yurdakul@kemerburgaz.edu.tr

S.M. Yalçın • Z. Gürhan-Canlı (✉)
Koç University, Istanbul, Turkey
e-mail: seda.muftugil@gmail.com; zcanli@ku.edu.tr

© Springer International Publishing AG 2017
T. Osburg, C. Lohrmann (eds.), *Sustainability in a Digital World*, CSR,
Sustainability, Ethics & Governance, DOI 10.1007/978-3-319-54603-2_19

the details of the inclusive business activity conducted by Turk Telekom and explain the key success factors of the project, including the strong collaboration and commitment between its partners. The chapter concludes with the business and social impact of the project through the discussions on the link between consumer trust, adoption of innovation and attitude change. These discussions also show how ICT can contribute to development efforts and reducing the market separations in the BOP markets.

The growing literature on win-win strategies to address low-income communities has spelt the optimal characteristics of a pro-poor innovation (see for example: Prahalad and Hart 2002). This literature identified various strategies for 'co-creation' or joint-value creation with the user community through non-traditional forms of collaboration (Brugmann and Prahalad 2007; Franceys and Weitz 2003; London et al. 2006). Turk Telekom's initiative provides an interesting case in that it presents an innovative approach to tackle a persisting problem in Turkey in a sustainable manner through its usage of technology. Moreover, the project shows a good understanding of the dynamic interplay between users' skills and abilities, social contexts and technological artefacts/applications, which is required to serve to BOP customers profitably (Lal Dey et al. 2013).

Inclusive growth (and by extension inclusive innovation) has been widely acknowledged as a goal of public and business policy (George et al. 2012). The proliferation of cases similar to Turk Telekom's initiative done by companies all over the world explicates that organizations can and do engage in social innovation activities to connect disenfranchised individuals and communities with opportunities that foster social and economic growth. As a result, inclusive growth diminishes trade-offs between growth and inequality because the poor become enfranchised as customers, employees, owners, suppliers, and community members (George et al. 2012). Abundant evidence shows that the efforts of private sector firms to engage the poorest "Base of the Pyramid" (BoP) households as consumers and producers— when successful—can result in significant improvements in the quality of life of the world's poor (Hammond et al. 2007; Marquez et al. 2010; Prahalad 2010; Rangan et al. 2007).

According to UNDP "inclusive business models include the poor on the demand side as clients and customers and on the supply side as employees, producers and business owners at various points in the value chain" (UNDP 2008: 14). Similarly, this term is used by the alliance between the World Business Council for Sustainable Development (WBCSD) and the Dutch development organization (SNV), to mean an "economically profitable, environmentally and socially responsible entrepreneurial initiative, which integrates low-income communities in its value chain for the mutual benefit of both the company and the community" (WBCSD and SNV 2010: 13). According to G20 Inclusive Business Framework, there are three approaches to conducting inclusive business: Inclusive business models, inclusive business activities, and social enterprise initiatives (G20 Inclusive Business Framework 2016: 5). In order to be successful at each approach, a high-level commitment and a long term support is required. According to this G20 classification, companies with an inclusive business model integrate the BOP into their core business operations. In this model "commercial viability of the business model is at the forefront for companies as they rely largely on commercial sources of financing for the business operations" (G20 Inclusive Business Framework 2016: 5). Inclusive business activities, just like inclusive business models, also include people at the BOP into companies' value chains. However most of the times, these activities are not central to the commercial viability of the company nor do the BOP constitute an influential art of the base of customers, suppliers or business partners. The last approach; namely social enterprises on the other hand are not designed to maximize their profits for redistribution. Most of the time, profits are reinvested back into the enterprise in order to fulfil and strengthen its social mission (G20 Inclusive Business Framework 2016: 5). In this scheme, "Life is Simpler with Internet" of Turk Telekom proves to be an inclusive business activity carried out by the current CSR team in charge at Turk Telekom.

2　Technology and Innovation in the Context of Inclusive Business

The post 2015 agenda on sustainable development has emphasized the role of science, technology and innovation in promoting sustainable development (Chandran et al. 2015; Dosi and Freeman 1988; Fagerberg et al. 2010). Increasingly innovation has been proposed as a fundamental ingredient for development (Dosi and Freeman 1988; Fagerberg et al. 2010; Freeman and Soete 1997) and how innovation emerges and diffuses under conditions of resource constraint within developing countries has become a topic of increasing interest in the academic literature (Lundvall et al. 2009). Especially, Information and Communication Technology (ICT) is believed to be capable of reducing particular types of market separations between consumers and producers at the BOP, and thus facilitating market development at BOP (Tarafdar and Singh 2011). However, the literature also cautions that the purchase or use of ICT products and services is, at least as much as it is an economic or technological act, a social act (Burrell 2008; Horst and Miller 2006). ICT products are carriers of personal, social, and cultural meaning, such as establishing a new place in society, fulling an aspiration for the future (Kuriyan et al. 2008). Therefore, as Anderson and Billou (2007) argue, because of the nature of the BOP, if companies want to enter this market, they have to consider the 4As of awareness, acceptability, affordability and availability. In other words, MNCs should make sure that poor people are aware of the existence of their product and the advantages it can offer, that they can afford to purchase the product and the product is available in their village if they decide to buy it (Tasavori et al. 2015). In sum, it can be concluded that necessary conditions to the successful adoption of an innovation in the BOP context includes a real need, compatibility of innovation with need, positive consumer perception of innovation value, and the use of change agents and accessibility to market in which the innovation is supplied (Simanis 2010). Turk Telekom's initiative addresses all these issues identified in the literature for adoption of internet among the poor.

3   Life Is Simple with Internet Initiative: Paradigm Shift at Turk Telekom

Türk Telekom Group is Turkey's integrated telecommunication and technology services provider. Turk Telekom Group Companies have a wide service network and product range in the fields of individual and corporate services. As of December 31, 2015, Turk Telekom has 12.9 million fixed access lines, 8 million broadband and 17.3 million mobile subscribers. Turk Telekom Group companies provide services in all 81 cities of Turkey with more than 34,000 employees. In January 2016, Turk Telekom unified its mobile, fixed voice, broadband and TV products and services under the single "Turk Telekom" brand.

TTNET, a former independent company in the Turk Telekom Group, was actually where the "Life is Simple" initiative was first born. It is now carried under Turk Telekom as the former brands dissolved under one brand as stated. TTNET, which was founded in 2006 as a communication and entertainment company joined the Business Call to Action in 2013 with a commitment to broaden internet access and internet literacy training and thus became the first Turkish company to join the Business Call to Action (BCtA). TTNET expanded internet access and educational opportunities to remote areas in Turkey since the inception of "Life is Simpler with Internet" project. The project aimed to increase the internet literacy rate in Turkey, providing capacity improvement support for using the e-Services offered by the public and private sectors, creating awareness on conscious use of the new media tools and providing information to the society on secure internet. The project introduced the online world to 12,000 people, who did not have the opportunity to use the internet before, with the cooperation of TTNET, United Nations Development Program (UNDP) and Habitat Development and Governance Association. Thus, it aimed a large audience who has not met with the online life yet. The project focused on middle age and above individuals, who were not internet literate and who needed basic information and skills, and aimed to facilitate their lives by supporting them to become internet literate. In more concrete terms, according to BCTA reports, the project, when started aimed to bring 250,000 people online to become regular Internet users in 2014, with a special focus on low-income consumers; provided up to 30,000 households with access to an online education platform; and delivered Internet literacy training to 12,000 disadvantaged people in 20 cities.

Ms. Gürsoy and Ms. Özgen revealed that the reasons of engagement in the project was multilayered and the idea of the project occurred at a time when they, as members of CSR department, were questioning the effectiveness and impact of their CSR projects, which were then mostly based on philanthropic ad hoc investments (generally consisted of building schools or making donations to schools in Anatolia) that were not related to the company's business. As a CSR department, they were not able to measure the social impact of these donations most of the time, which rendered their activities invisible for the top management whose orientation was towards profit maximization. They were quick to realize that nonstrategic social responsibility projects were not being effective and sustainable for neither the beneficiaries nor the company itself. The research they conducted about the visibility of their CSR efforts in Anatolia with respect to philanthropic donations such as building schools turned out to be rather invisible even for the local communities where the schools were present.

Due to their previous experience from conducting philanthropic CSR activities, the CSR department came up with a new idea that might increase the traffic of telephones—which was simultaneously on the business agenda of the management team. The idea was to open a phone library for the visually handicapped people through audio books. The idea was very much liked by the top management and this project proved to be successful in increasing the reputation of the company and at the same time increasing the phone traffic as they predicted. Ms. Özgen said "They (top managers) were grateful. Because of this experience, the management team started to think the social responsibility could be done by supporting our business." By this time, Ms. Özgen and Ms. Gürsoy were already knowledgeable about concepts of inclusive business and struggling to show its possible merits to the top management. They knew that this project was an inclusive business activity but they said that by then "the term inclusive business was still within their department" and the term was not frequently used by the top management.

In 2012, the size of the market appeared to be not growing, fixed at six million internet users. A call conference was made inside the company, inviting different departments to think about what might be done to grow the market. Ms. Özgen and Ms. Gürsoy were also invited to these sessions and they thought that non-users were an important group that could "actually be covered through corporate social responsibility". They thought; "If we design this project with business objectives, we can share the project with the management and get their support." Their strategy proved right and that was how they convinced the managers to design the project with an inclusive business lens. "We talked about our vision of social responsibility; tried to design a project in such a way that it is related to our business." Backed by the top management, Ms. Özgen and Ms. Gürsoy had a full mandate to carry on their project proposals and materialize it through financial and motivational support.

The literature indicates that level of commitment and its longevity are key determinants of success in inclusive business, specifically due to the long-term returns of these practices. In terms of corporate commitment for responsible business four components are proposed as key factors for long-term engagement and holistic development. These are respectively; top management and supplier support (Hoejmose et al. 2012), shared values and objectives within the organization, a robust ethical foundation (Del Baldo 2013), and increasing leadership and stakeholder activity and motivation (Duran-Encalada and Paucar-Caceres 2012). In the Turk Telekom case, all of these four components have been existent albeit with varying degrees.

Collaboration

One of the major reasons for the need to form collaborations is that partner heterogeneity affects the environmental outcomes as these partnerships tend to follow more proactive strategies compared to inter-firm alliances (Lin 2012). Collaborating with the government is specifically important for the BOP markets mainly due to the scope of the problems and capacity gaps. Public–private partnerships are suggested to increase the social impact of the responsible initiatives in which the role of the partner is policy implementation, while the role of government is policy development (Galea and McKee 2014).

Similarly, Blok, Sjauw-Koen-Fa and Omta (2013) call for attention to the importance of collaboration between for profit and not-for-profit organizations, customers and all stakeholders to achieve both economic and social goals at the BOP markets. Partnership of local entrepreneurs and development partners such as civil society groups, the government and corporations may suggest an opportunity to create a fortune with the BOP, rather than at the BOP (Calton et al. 2013). Decentralized stakeholder networks, global action networks and a more human focused and inclusive perspective are required to reach the expected benefits of collaboration (Calton et al. 2013).

"Life is Simple with Internet" initiative was designed as a collaborative one from the beginning encapsulating an NGO, namely Habitat, and International agency, UNDP and government. "We talked to UNDP, Habitat and Ministry of Development and it is through this collaboration that the content and the design of the project emerged," said Ms. Özgen. Previous research lays out various reasons for collaboration among different stakeholders such as social capital for development (Gatune 2010, Fisher et al. 2009), social license and corporate—community involvement (Idemudia 2009a, b). Collaboration between central and local governments, local partnerships and change agents is proposed to secure sustainable development (Nielsen and Thomsen 2011), to ensure environmental management (Cheung et al. 2009), and to increase the expected impact of these initiatives (Sanneh et al. 2014).

A major stream of research on collaborations focus on cross-sector partnerships especially in the BOP markets, which help companies in overcoming the problems faced in unfamiliar BOP markets (Schuster and Holtbrügge 2014). Furthermore, partners from multiple sectors may help in addressing the problem of institutional gaps in these markets (Rivera-Santos et al. 2012). For example, partnerships with NGOs can help in eliminating the contextual problems MNCs face in BOP markets, thanks to NGOs knowledge of the context, and their social embeddedness (Webb et al. 2010).

Not only Ms. Gürsoy and Ms. Özgen were aware of the fact that they needed to work with a local NGO present in the underdeveloped cities targeted, but they were also aware that this cooperation had to be of a certain quality. "We had great advantages by cooperating with Habitat" they stated. Habitat was the civil society organization that managed and found volunteers who actually gave the internet literacy trainings in the project. Instead of spending time and effort to do this by themselves, project leaders clearly opted out to work with Habitat, who chose volunteers, most of the time local young people who were aware of the cultural context of the cities where the trainings took place. CSR team at Turk Telekom was clearly attuned to the literature which stated that innovations that are social in focus should match the members of that society, its context, and the environment (Reynoso et al. 2015). Their close connection with the local youth, and their constant feedback loop made this process rather effective. We were told that these young people, who were most of the time from that particular city where the training was given, tried different methods to attract people; especially for older women who were hesitant to go outside home. In some cases, these volunteers came up with innovative solutions and carried out the trainings in the homes of women where they were gathering to socialize and thus changed the place of training, thinking this might increase the efficiency of the program. In other cases, these young people were trying to fight with "prejudices and biases about internet usage's potential to destroy marriages" prevalent in a specific region, in cunning ways so that they ensured attendance to the trainings. Cultural barriers like these, according to the CSR team, could be better addressed through local partners that have experience on these issues.

Hahn and Gold (2014) suggests that, generally to initiate an alliance one partner must have the ability to identify potential partners with synergetic potential. This ability according to these scholars, are influenced by several factors. These factors can be enumerated as prior alliance experience that helps to spot potential partners and their resources. This also indicates that the companies should have the ability to screen potential partners for their capabilities and resources (Liao et al. 2008). In addition, "an information-rich position in the socioeconomic network ensures superior access to reliable information about (potential) partners that makes fitting partnerships more probable and their exploitation more effective" (Hahn and Gold 2014:1323).

Our in-depth interview indicated that "Life is Simple with Internet" initiative clearly carried the elements of these several factors, which made this initiative a success. The literature on inclusive business suggests that the success of partnerships among the MNCs, NGOs, and the poor are contingent on establishing trust through dense networks, partnering with well-reputed NGOs and MNCs, and establishing both voluntary and enforceable codes of conduct among the partners (Shivarajan and Srinivasan 2013). Collaboration among businesses, governments, consumers, civil society and financial sector is needed to reach mutual goals and to minimize social and environmental costs through holistic solutions (Al-Tabbaa et al. 2014; German et al. 2011; Rotter et al. 2014; Vidal-Leon 2013).

4 Business and Social Impact of Life Is Simple with Internet Project

As emphasized in post-2015 agenda, science, technology and innovation have a key role in sustainable development (Fagerberg et al. 2010; Chandran et al. 2015). Even though the households that have access to internet have risen from 60.2% (2014) to 69.5% (2015), 29.5% of the population—and especially the economically disadvantaged groups—still lack internet connection in Turkey. Considering the proliferation rate of internet technologies, BOP is a market to be tapped in with very low levels of saturation. Previous studies show the willingness of disadvantaged groups to accept ICT thanks to the increase in penetration rates after successful projects customized for these groups (Urquhart et al. 2008). As a first step to create this market, TTNET delivered internet literacy training to 12,000 disadvantaged people. However, as the literature points out, long term efforts and iterative activities are needed to ensure the adoption of these technologies (Venkatesh and Bala 2008).

Inclusive business activities are focused on creating a win-win situation for all stakeholders. Therefore, by definition, these activities aim to create both social and business impact, which is fundamental for their sustainability. Considering the opportunities that may arise through internet access in the base of the pyramid (BOP) markets, TTNET designed the project as an inclusive business activity primarily focusing on the social impact, as explained by Ms. Gürsoy:

> In fact, as the corporate social responsibility department, our first driver was the social impact. After all, we are communicators and we need to contribute to the reputation of the company. If you are a communicator, your primary job is to create a project, which can appear in the news, which can contribute to reputation. But social responsibility projects shrink in time. The reason for getting closer to inclusive business was this: we can plan the project with the business impact and can show this to the top management. We no longer say to the top management that "this has been on the news, and this number of people watched it". We see in the impact report that this project created an attitude change, people started to say that the internet worth the money they pay. We also think in line with this because we are able to get approval easily. We can get the budget approval as the top management is easily convinced this way.

Social impact provided their department with the required support from within the organization. The CSR team also states that they did not want to have a commercial focus in this project as the core aim is to create attitude change; even though, in the long run, they have an expectation of market development as a business outcome. This outcome requires systematic, long-term efforts due to the fact that neither attitude change nor the behavioral change that follows comes quickly. The team thinks that the motivation for this project was equally divided between social and business impact (market development on one side, and social impact on the other).

Another reason for the business impact to be in the second place in their agenda was the barriers for marketing TTNET products during and after the project. The marketing team developed a new, affordable offering customized according to the needs of the BOP market. However, they were not able to promote this product, as one of the main partners of the project did not approve its promotion as it is against the rules of fair competition. On the other hand, TTNET had a 2-year free offering of a low quota internet connection for the first time users, which they wanted to offer to the participants of the training program. However, due to the same concern, this offer was not approved either. This was not perceived as a major problem by the company as the primary aim was not an increase in sales, but to create an attitude change as a first step to be taken in their efforts for tapping into BOP market. Furthermore, Ms. Gürsoy states that as per their previous experiences, providing the free package did not lead to a change in consumers' attitudes, while the survey conducted after the training shows a positive change in favor of internet adoption.

Still, to overcome this barrier, they created an offering including internet connection and a limited-term subscription to an online training platform for children, and priced it accordingly (9.90 TL per month). However, the ICT sector is regulated in Turkey and the Information and Communication Technologies Authority (BTK) did not allow TTNET to offer this package only in regions with development priority, with the concern of equal opportunity. Consequently, TTNET provided this special offer to the whole country, but promoted it only in these regions through customized advertising campaigns in these cities.

Due to these reasons, and TTNET's focus on the social impact, it was not possible to track the business impact of the project in terms of its contribution to sales, brand preference or brand image. On the other hand, social impact was measured and reported, which also gave the company important clues about the potential business impact of the project in the forthcoming years.

The data clearly reveals the attitude change; however, there is no data at hand to see whether this change in consumers' attitudes have led to a behavioral change (such as subscriptions or effective use of the internet after the training). On the other hand, TTNET tracked that the trainees were using internet services such as e-government services, social media and other tools during the training, which can be considered as an indicator of their future use.

5 Technology Adoption at the BOP Markets and the Design of the Project

Co-creation with the community becomes a key success factor for the offerings in the BOP market (Brugmann and Prahalad 2007; Hart 2005; London 2008). Consequently, one of the major steps to be taken was to have a solid knowledge about the context. TTNET solved this problem through their volunteers, as they were citizens themselves who are aware of the potential cultural, social and economic problems, and able to propose innovative and effective ways to solve them. Furthermore, the training program was formulated with a consideration of the barriers for internet use in the low income communities which were identified through previous research of Turk Telekom and TUIK (Turkish Statistical Institute). The project was designed according to the data received from TUIK about the attitudes of non-users towards the internet. In addition, the marketing department conducted a research on non-users, which showed that the basic reasons for not using internet was its price, followed by lacking a device to connect to the internet, and problems associated with mistrust in the technology.

According to the well-known technology acceptance model (TAM) of Davis (1989), perceived usefulness and perceived ease of use are two main drivers of individual's adoption intention and usage of technology. Venkatesh and Davis (2000) extended the TAM model and argued that social influence processes (subjective norm, voluntariness and image) and cognitive instrumental processes (job relevance, output quality, result demonstrability and perceived ease of use) are factors that significantly influence the acceptance of technology. On the other hand, Roger's (2003) diffusion theory reveals that users' acceptance of an innovation is influenced by their perception of its relative advantage, compatibility, complexity, trialability and observability. The innovation adoption increases if the innovation provides a relative advantage, is compatible with users' existing experiences and values, easy to use, give opportunities for trial and if the benefits of adopting the innovation is easily observable. Considering the important role of ICT in development, further studies have been conducted on technology adoption at the BOP markets. In their study on mobile banking, Ismail and Masinge (2011) argues that perceived usefulness, perceived ease of use, perceived cost and consumer trust were effective in the adoption of mobile banking.

According to TUIK data, low literacy, training and income levels contributes to the perception that internet is an unsafe, addictive and harmful tool. In line with the TAM model, the data reveal that:

- Perceived usefulness: 59.5% of the non-users think that they do not need internet.
- Perceived ease of use: 44.7% does not have internet as they do not know how to use it.
- Perceived cost: 38.5% thinks that the price is too high while 36.5% thinks that they do not have devices to connect to the internet as the prices of these devices are too high for them.

With a similar set of findings, the research conducted by TTNET with the participants before the training shows that:

- Perceived cost: 70% thinks that having an internet connection at home will be too costly.

- Consumer trust: 52% thinks that internet is not safe, while 41% state that they are afraid of using internet. On the other hand, 61% of the participants think that internet can pose a threat to their marriage and 71% thinks that it is dangerous for their children. Seventy nine percent believes that internet weakens the communication between family members.

In light of these results (which are congruent with the findings of previous studies), TTNET customized their training programs accordingly in order to create a positive change in the attitudes of the participants towards internet use. Social impact of the project was measured through post-participation surveys, revealing the positive change in the attitudes. The percentage of participants who think that:

- Having an internet connection will be too costly decreased from 70 to 60%.
- Internet is not safe decreased from 52 to 37%.
- Afraid of using internet decreased from 41 to 29%.
- Internet can pose a threat to marriages decreased from 61 to 45%.
- Internet is dangerous for the children decreased from 71 to 52%.
- Internet weakens the communication between family members decreased from 79 to 55%.

Considering the perceived usefulness, participants revealed in the post-tests that internet is a necessity of the modern life, provides them with opportunities to connect to their relatives and loved ones, and help them in bridging the generation gap between themselves and their children. After the training, 65% of the participants started using e-government services, 58% of the participants started using online communication tools, and 56% started using social media and news websites. Participants' knowledge about the tools available to their use through internet (such as e-government services, online banking services and online shopping) increased for all types of services. Furthermore, the training also contributed to democratization as it gives the opportunity to become more participative and to communicate with the government and other authorities. On the other hand, one of the major business impacts of the activity was that, participants' attitude towards the cost of internet has changed as they reveal that the money they will be paying for the internet will worth it.

6   Discussion

As the data reveal, the training contributes to the evaluations of the participants regarding the main drivers of technology adoption, namely, perceived usefulness, perceived ease of use, perceived cost, and most importantly, consumer trust. Consequently, through the attitude change, the training contributes to the adoption of ICT, which shows us the significance of the social impact, as well as the potential business impact of the activity.

Even though TTNET did not measure the business outcome due to their focus on the social impact, it is clear that the training program contributes to their efforts of market development. Improved levels of trust in ICT may lead to the adoption of the technology which can contribute to the market share and future profitability of the company. On the other hand, in consideration of the reciprocity principle, it is likely that the participants may develop a positive attitude towards the brand, affecting their brand preference and brand loyalty. Finally, considering the very high levels of satisfaction among participants (93%), the training may also enhance the brand image.

Lack of access to goods and services due to problems related to accessibility and affordability is a major problem in BOP markets (Karnani 2007; Prahalad 2005). ICT can play a key role in reducing these market separations between consumers and producers in the BOP markets (Tarafdar and Singh 2011). Closing these gaps lead to market development. According to Bartels's theory of market separations (1968), there are four types of separations in BOP markets: spatial (geographical distance between buyer and seller), temporal (time difference between production and consumption), informational (informational asymmetry between producers and consumers in terms of products, market conditions etc.) and financial (lack of financial resources to purchase).

As argued in the previous studies, and supported by the data from the social impact measurement of this project, ICT contributes to bridging the separation (Tarafdar and Singh 2011). In this case, easy access to online products and services reduces the temporal and spatial separation. Financial separation is reduced through making products and services more affordable to the BOP consumer. Consumers could search for alternative products and services and find the best offers available in the market through the internet. Therefore, ease of access to information through ICT reduces both the informational separation and financial separation. Finally, it is possible to argue that internet may also reduce the financial separation due to ease of access to skill development and job opportunities.

Previous studies also emphasize that technology adoption does not guarantee the effective use of technology (Dhir et al. 2012; Walsham 2010) as it needs to be appropriated through continuous interactions between users and technological applications (Dey et al. 2013). Furthermore, technology appropriation is context dependent and influenced by macro environmental factors as well as individual abilities (Dey et al. 2013). Considering the fruitful outcomes of this project, and business and social impact that could be created through the proliferation of ICT in BOP markets, we recommend companies to employ long-term, iterative efforts of awareness creation, knowledge and skill development and technology appropriation to reap the desired benefits.

References

Al-Tabbaa O, Leach D, March J (2014) Collaboration between nonprofit and business sectors: a framework to guide strategy development for nonprofit organizations. Voluntas 25(3):657–678

Anderson J, Billou N (2007) Serving the world's poor: innovation at the base of the economic pyramid. J Bus Strategy 28(2):14–21

Bartels R (1968) The general theory of marketing. J Mark 32(1):29–33

Blok V, Sjauw-Koen-Fa A, Omta O (2013) Effective stakeholder involvement at the base of the pyramid: the case of Rabobank. Int Food Agribus Man 16(A):39–44

Brugmann J, Prahalad CK (2007) Cocreating business's new social compact. Harv Bus Rev 85 (2):80–90

Burrell J (2008) Problematic empowerment: west African internet scams as strategic misrepresentation. Inform Technol Int Dev 4(4):15–30

Calton JM, Werhane PH, Hartman LP, Bevan D (2013) Building partnerships to create social and economic value at the base of the global development pyramid. J Bus Ethics 117(4):721–733

Chandran VGR, Kwee NB, Yuan WC, Kanagasundaram T (2015) Science, technology and innovation for inclusive development: reorganizing the national and regional systems of innovation. Tech Monitor, Jan–Mar:14–19

Cheung DKK, Welford RJ, Hills PR (2009) CSR and the environment: business supply chain partnerships in Hong Kong and PRDR, China. Corp Soc Resp Env Ma 6(5):250–263

Davis FD (1989) Perceived usefulness, perceived ease of use and user acceptance of information technology. MIS Quart 13(3):319–340

Del Baldo M (2013) Corporate social responsibility, human resource management and corporate family responsability. When a company is "The best place to work" Elica group, the hi-fi company. Ekonomska istraživanja – Econ Res Special Issue 26:201–224

Dey BL, Binsardi B, Prendergast R, Saren M (2013) A qualitative enquiry into the appropriation of mobile telephony at the bottom of the pyramid. Int Market Rev 30(4):297–322

Dhir A, Moukadem I, Jere N, Kaur P, Kujala S, Yla-Jaaski A (2012) Ethnographic examination of studying information sharing practices in rural South Africa. In: Paper presented at the 5th international conference on advances in computer-human interactions. www.academia.edu/1405296/Ethnographic_Examination_for_Studying_Information_Sharing_Practices_in_Rural_South_Africa. Accessed 20 Jan 2013

Dosi G, Freeman C (1988) Technical change and economic theory. In: Dosi G, Freeman C, Nelson R, Silverberg G, Soete LL (eds) Laboratory of economics and management (LEM). Sant'Anna School of Advanced Studies, Pisa

Duran-Encalada JA, Paucar-Caceres A (2012) A system dynamics sustainable business model for Petroleos Mexicanos (Pemex): case based on the global reporting initiative. J Oper Res Soc 63 (8):1065–1078

Fagerberg J, Srholec M, Verspagen B (2010) Innovation and economic development. In: Hall B, Rosenberg N (eds) Handbook of the economics of innovation, vol II, North Holland, pp 833–872

Fisher K, Geenen J, Jurcevic M, McClintock K, Davis G (2009) Applying asset-based community development as a strategy for CSR: a Canadian perspective on a win–win for stakeholders and SMEs. Bus Ethics 18(1):66–82

Franceys R, Weitz A (2003) Public private partnerships in infrastructure for the poor. J Int Dev 15 (8):1083–1098

Freeman C, Soete L (1997) The economics of industrial innovation, 3rd edn. MIT Press, Cambridge, MA

G20 Inclusive Business Framework (2016) http://www.ifc.org/wps/wcm/connect/f0784d004a9b1f2ea5f0ed9c54e94b00/Attachment+G+-+G20+Inclusive+Business+Framework_Final.pdf?MOD=AJPERES. Accessed 20 Dec 2015

Galea G, McKee M (2014) Public-private partnerships with large corporations: setting the ground rules for better health. Health Policy 115(2–3):138–140

Gatune J (2010) Africa's development beyond aid: getting out of the box. Ann Am Acad Pol Soc Sci 632:103–120

George G, McGahan AM, Prabhu J (2012) Innovation for inclusive growth: towards a theoretical framework and a research agenda. J Manage Stud 49(4):661–683

German L, Schoneveld GC, Pacheco P (2011) Local social and environmental impacts of biofuels: global comparative assessment and implications for governance. Ecol Soc 16(4):29–43

Hahn R, Gold S (2014) Resources and governance in "base of the pyramid"-partnerships: assessing collaborations between businesses and non-business actors. J Bus Res 67 (7):1321–1333

Hammond A, Kramer WJ, Tran J, Katz R, Walker J (2007) The next 4 billion market size and business strategy at the base of the pyramid. World Resources Institute, Washington, DC

Hart SL (2005) Capitalism at the crossroads: the unlimited business opportunities in serving the world's most difficult problems. Wharton School Publishing, Upper Saddle River

Hoejmose S, Brammer S, Millington A (2012) "Green" supply chain management: the role of trust and top management in B2B and B2C markets. Ind Market Manag 4:609–620

Horst HA, Miller D (2006) The cell phone: an anthropology of communication. Berg Publishers, Oxford

Idemudia U (2009a) Assessing corporate-community involvement strategies in the Nigerian oil industry: an empirical analysis. Resour Policy 34(3):133–141

Idemudia U (2009b) Oil extraction and poverty reduction in the Niger delta: a critical examination of partnership initiatives. J Bus Ethics 90:91–116

Ismail T, Masinge K (2011) Mobile banking: innovation for the poor. United Nations University, UNU-MERIT working paper series. http://collections.unu.edu/eserv/UNU:419/wp2011-074. pdf. Accessed 14 Feb 2016

Karnani A (2007) The mirage of marketing to the bottom of the pyramid: how the private sector can help alleviate poverty. Calif Manage Rev 49(4):90–111

Kuriyan R, Ray I, Toyama K (2008) Information and communication technologies for development: the bottom of the pyramid model in practice. Inform Soc 24(2):93–104

Liao SH, Chang WJ, Lee CC (2008) Mining marketing maps for business alliances. Expert Syst Appl 35(3):1338–1350

Lin HY (2012) Cross-sector alliances for corporate social responsibility partner heterogeneity moderates environmental strategy outcomes. J Bus Ethics 110(2):219–229

London T (2008) The base-of-the-pyramid perspective: a new approach to poverty alleviation. Acad Manage Proc 1:1–6

London T, Rondinelli DA, O'Neill H (2006) Strange bedfellows: alliances between corporations and nonprofits. In: Shenkar O, Reuer JJ (eds) Handbook of strategic alliances. Sage, Thousand Oaks

Lundvall B, Vang J, Joseph K, Chaminade C (2009) Bridging innovation system research and development studies: challenges and research opportunities. In: 7th Globelics conference, Senegal

Marquez P, Reficco E, Berger G (2010) Socially inclusive business: engaging the poor through market initiatives in iberoamerica David Rockefeller center for Latin American Studies. Harvard University Press, Cambridge, MA

Nielsen AE, Thomsen C (2011) Sustainable development: the role of network communication. Corp Soc Resp Env Ma 18(1):1–10

Prahalad CK (2005) The fortune at the bottom of the pyramid: eradicating poverty through profits. Wharton School Publishing, Upper Saddle River

Prahalad CK (2010) The fortune at the bottom of the pyramid: eradicating poverty through profits. Pearson Education, Upper Saddle River

Prahalad CK, Hart SL (2002) The fortune at the bottom of the pyramid. Strategy Bus 26:1–14

Rangan K, Quelch J, Herrero G, Barton B (2007) Business solutions for the global poor: creating social and economic value. Jossey-Bass, San Francisco

Reynoso J, Kandampully J, Xiucheng F, Paulose H (2015) Learning from socially driven service innovation in emerging economies. J Serv Manage 26(1):156–176

Rivera-Santos M, Rufin C, Kolk A (2012) Bridging the institutional divide: partnerships in subsistence markets. J Bus Res 65(12):1721–1727

Rogers EM (2003) Diffusion of innovations, 5th edn. Free Press, New York

Rotter JP, Airike PE, Mark-Herbert C (2014) Exploring political corporate social responsibility in global supply chains. J Bus Ethics 125(4):581–599

Sanneh ES, Hu AH, Njai M, Ceesay OM, Manjang B (2014) Making basic health care accessible to rural communities: a case study of Kiang West district in rural Gambia. Public Health Nurs 31 (2):126–133

Schuster RT, Holtbrügge D (2014) Benefits of cross-sector partnerships in markets at the base of the pyramid. Bus Strat Environ 23(3):188–203

Shivarajan S, Srinivasan A (2013) The poor as suppliers of intellectual property: a social network approach to sustainable poverty alleviation. Bus Ethics Q 23(3):381–406

Simanis E (2010) Needs, needs everywhere, but not a BoP market to tap. In: London T, Hart S (eds) Next generation business strategies for the base of the pyramid: new approaches for building mutual value. FT Press, Upper Saddle River

Tarafdar M, Singh R (2011) A market separations perspective to analyze the role of ICT in development at the bottom of the pyramid. In: Proceedings of SIG GlobDev 4th annual workshop, Shanghai, 3 Dec 2011. https://www.researchgate.net/profile/Ramendra_Singh3/pub lication/265351953_A_Market_Separations_Perspective_to_Analyze_the_Role_of_ICT_in_ Development_at_the_Bottom_of_the_Pyramid_A_Market_Separations_Perspective_to_Ana lyze_the_Role_of_ICT_in_Development_at_the_Bottom_of_the_Pyramid/links/ 55b7068808aec0e5f43803b6.pdf. Accessed 12 Feb 2016

Tasavori M, Zaefarian R, Ghauri PN (2015) The creation view of opportunities at the base of the pyramid. Entrep Region Dev 27(1–2):106–126

UNDP (2008) Creating value for all: strategies for doing business with the poor.http://www.rw. undp.org/content/dam/rwanda/docs/povred/RW_rp_Creating_Value_for_All_Doing_Business_ with_the_Poor.pdf. Accessed 20 Aug 2015

Urquhart C, Liyanage S, Kah MM (2008) ICTs and poverty reduction: a social capital and knowledge perspective. J Inform Technol 23(3):203–213

Venkatesh V, Bala H (2008) Technology acceptance model 3 and a research agenda on interventions. Decis Sci 39(2):273–315

Venkatesh V, Davis FD (2000) A theoretical extension of the technology acceptance model: four longitudinal field studies. Manag Sci 46(2):186–204

Vidal-Leon C (2013) Corporate social responsibility, human rights, and the world trade organization. J Int Econ Law 16(4):893–920

Walsham G (2010) ICTs for the broader development of India: an analysis of the literature. Electron J Inform Syst Dev Countries 41(4):1–20

WBCSD and SNV (2010) Inclusive business: creating value in Latin America. WBCSD and SNV, Geneve and The Hague. http://www.wbcsd.org/Pages/EDocument/EDocumentDetails.aspx? ID=43&NoSearchContextKey=true. Accessed 10 July 2015

Webb JW, Kistruck GM, Ireland RD, Ketchen DJ (2010) The entrepreneurship process in base of the pyramid markets: the case of multinational enterprise/nongovernment organization alliances. Entrep Theory Pract 34(3):555–581

"Down the Yellow Brick Road": Challenging the Existing Business Models

Amira Dotan, Yossi Rahamim, and Anat Even-Chen

1 Introduction

The traditional academic world is conservative, some may even argue—an ortho-dox one. However, in recent years, ambiance of transformation emerges in varied forms, mostly due to free access to information throughout the open electronic media, all calling for a change. We, as part of the Israeli academic field representing Corporate Social Responsibility (CSR), see this change as a venue which enables us to create new opportunities, through which we offer our students to be part of the process to form a new business model. The four major changes perceived, in our view, as the basis of new management concepts are as follows:

- The "small globe"—the understanding that our globe, shared by all humanity and all living creatures, is a small planet and not inexhaustible, and therefore we affect one another even when we think our actions are local;
- Quick development of social technology, which enables faster and tighter contacts among people all over the world;
- Realization of the limited resources of natural and environmental assets;
- Corporate global power which exceeds governments' power and therefore might also harm more than governments can.

In a nutshell, CSR view Corporations as "Citizens" of society and of the world, hence, responsible to all the stakeholders who are influenced by them or are part of

A. Dotan (✉) • A. Even-Chen
Academic Corporate Social Responsibility Center, School of Business Administration,
The College of Management Academic Studies (COMAS), Rishon LeZion, Israel
e-mail: damira@colman.ac.il

Y. Rahamim
Haim Striks School of Law, The College of Management Academic Studies (COMAS),
Rishon LeZion, Israel

© Springer International Publishing AG 2017
T. Osburg, C. Lohrmann (eds.), *Sustainability in a Digital World*, CSR,
Sustainability, Ethics & Governance, DOI 10.1007/978-3-319-54603-2_20

them. This, by itself, means a major change and a call for implementing processes. The term "Corporate Citizenship" (CC) is well known in CSR literature, for example, Matten and Crane (2005) in their article "*Corporate Citizenship: Toward An Extended Theoretical Conceptualization*"; a new journal dedicated specifically to this topic—The Journal of Corporate Citizenship, published by Greenleaf Publishing; and many more.

2 We and the New Concept

We believe that each one of us, managers and change agents, should implement the CSR values in our own yard and become leaders and example for others. Therefore, in this short essay, we will openly share our experience as a leading part of the School of Business Administration (SBA) of the College of Management Academic Studies (COMAS). We look at our place as a citizen in a corporate, in other words, we see students as well as faculty members, as our main stakeholders.

The Academic CSR Center within SBA is the pillar of CSR in COMAS, ergo we have decided to act as well as teach and give our students the finest tools in order to become responsive and work in teams, principals which are the core of nowadays management. The Center perceives its responsibility in guiding and motivating the students to be the managers of tomorrow and think "out of the box", by inventing new ways to lead as both persons and managers.

While the concepts and logic behind CSR are intuitively simple, implementing them, is rather challenging. Especially in our "here & now" world where things are instant and consequences are immediate, while attention to changes is a process which needs long term thinking. We took active as well as ideological part in a change process which began almost 4 years ago and is continuing to be dynamic and a rolling ball. The essay will reflect the various difficulties, obstacles and successes, pointing out the unique relationship between—physical and conceptual issues in changing the management model to an innovative academia.

3 Implementation

Our world is complex, therefore a transverse conceptual change is essential. The strategic thinking has been modified from linear thinking to a multifaceted one. "Top" manager/responsible person have to manage/conduct a transparent and open operation interactively. As an example of setting a tone, the SBA's Dean formed a School management led by a multidisciplinary committee of seven members, which promotes various ways of thinking (age, gender, status, organizational history keeper, law, financial and CSR). They meet once a week to conduct, design, learn, decide and deal with emerging problems. Three main pillars were built:

Learning HUBS

Change, according to our experience, needs to be seen and felt. Ambiance is crucial, therefore "old ordinary" classes were physically transformed to well equipped, welcoming open spaces for peer learning, guidance and assistance. This significant change creates a convenient, informal and intimate atmosphere which formed an added value for both the students and the academic stuff. Furthermore, it encourages studying together based on a dynamic dialogue, immediate assistance, openness and cooperation (see Fig. 1).

Mentors

New management needs a different way of "listening". To achieve it, young academic staff (educators) was chosen to mentor all first year BA students. Mentoring is consisted of various spheres of life—in campus and life experience. The mentors are the bridge to the Chairs of Academic Departments in order to ease the students' integration.

Access to Knowledge

One of our main guiding lights is that knowledge must be available to everyone and everywhere. In SBA we believe that a student should have all that is necessary to learn. We acknowledge the fact that life has its own withholding circumstances such as army reserves duty, pregnancy, illness etc. hence every semester we film and record a variety of courses, in the BA and the MBA programs. All of them are available on our YouTube channel. This way students can take their classes, find a substitute lecture if they missed one and avoid re-sit classes.

Fig. 1 Learning hubs

In addition, these recordings are open to the general public, enabling anyone with internet connection to watch, learn and widen their horizons. Some of the lecturers are key business persons or finance managers from the business community in Israel and worldwide, exposing students to maximum updated and personal information, that among other things enable vivid connection between the academic and practical methods.

4 Challenges

Our discussed perception of an open, flexible approach makes everyone feel as part of the management. The integration of different ways of thinking, abilities and mistakes form an eco-system that cope better with the various challenges (despite these challenges) and needs. The three actual changes listed above, echo the school principals—Be active, be partner and be initiative. They also accelerate learning excellence, hard work awareness and tools for acting as a citizen. While within SBA this new thinking and out of the box implementation is on action and fruitful, we face complications in the widen scope. Managing and dialoguing with all COMAS stakeholders can be quite a challenge. For example:

Financial Aspect
Exchange the use of white papers to 100% recycled paper (more expensive) and implement it as a "must have" product.

Sustainability Aspect
There is a clash between longing to build a wellbeing campus, which offer students and employees the enjoyment of healthier study and work environment, and COMAS refusal to direct the academic stuff to reuse study materials. This clash angered one of our students (who experienced the SBA policy) so much, therefore, with our assistance, he initiated and built a SOCIAL LIBRARY, enabling students to carry forward and reuse of the learning material. That way the students save both money and paper (see Fig. 2).

Transparency Aspect
Transparency is a major obstacle in the new manager's life that is still looking for a way to build itself. The clash between what should be open and to whom mirrors aspects of responsibility and ethics. We believe that's haring and transparency are one of the milestones to a sustainable organization, led by a flexible and well-adjusted management. To create the optimal transparency, it requires full collaboration with the stakeholders and ability to handle failures and success all together, in an honest and genuine approach.

Contractor Employees Aspect
Although employees are considered as core, COMAS still hires contractor employees. The change is still in its "first steps". The positive aspect is that the

Fig. 2 Social library

demand for a change was originated by students, who put the issue on COMAS' management table, with the assistance of knowledgeable faculty's members.

These examples show that creating a new business model is a long term process which clashes with material obstacles like finance, and overcome behavior and psychological barriers. We believe that we are now in the beginning of the blossom period where the "Greenwash" concept is no longer an option and the change is initiated by our students, who are the "future" managers. We are proud to be the ones to plant the seeds.

5   The Legal Basis for Considering Stakeholders

From Israeli Law perspective, the change in companies' liability towards their stakeholders was actually expressed in 2000. The new Company Law enacted by the Knesset (the Israeli Parliament) became valid. For the first time the Israeli Law anchored the companies' liability to take into consideration the interests of their employees, creditors, and the public in general.

Article 11(a) of the Israeli Companies Law states as follows:

The company goal shall be to operate in accordance with business considerations in realizing its profits, and within the scope of such considerations, the interests of its creditors, its employees, and the public may be taken into account. In many ways, even comparing to corporate laws of other western countries, this article is a substantial innovation in the Israeli law.

Firstly, for the first time the Israeli legislator explicitly referred to the question of the purpose of the company.

Until 2000, the Israeli Supreme Court was the only one to set in its ruling guidelines the proper liability to be imposed upon the companies, their officers and controlling shareholders. However, the Israeli Supreme Court was extremely cautious. The validation of the new Company Law anchored these rulings in the written law, and it paved the way to keep the companies' stakeholders interests.

Secondly, although the corporate laws in many countries in Europe and in the United States include similar instructions allowing the directors and officers in the company to take into consideration the interests of the stakeholders (as part of their duty of conduct), only the Israeli Company Law dedicated a specific article titled "Company Goal", in the chapter of the law preamble.

Therefore in the Israeli law, the article of the Company Goal serves as sort of a compass for the companies, and the modern perception embodied in this article passes like a thread throughout all other instructions of the Company Law. The perception that the company goal is no longer the benefit of the shareholders, but of the company itself (according to the best benefit of its members as a whole).

For the Academic CSR Center, the Company Goal article is the legal basis for advancing the managerial culture of companies to show liability towards the stakeholders, and base on that, to encourage and create a new business model. The Company Goal article may give a tailwind to a progressive perception according which the interests of the company's employees, credit suppliers and the public in general are seriously taken into consideration, in addition to its business considerations. This consideration is not taken instead of the shareholders' interests, but in addition to them, as a worthy and more efficient tool to lead the company to prosperity and long term stability.

Despite the aforesaid innovation in the Israeli Company Law, and its conceptual importance for the advancement of a managerial culture that fits the twenty-first century, as mentioned supra, this article is not free of criticism. The first criticism that is raised pertains to the declarative wording of the article, and to its vagueness. The article of the Company Goal was a priori drafted as a declarative article that did not set an internal mechanism for its implementation.

For example, the article does not set clear standards for cases in which one needs to take into consideration the stakeholders' interests; the article does not explicitly state who is the organ in the company (director or other officer) that carries the duty to act according to the instructions of the article; and the article does not grant to the stakeholders a cause of action when the company did not take their interests into consideration. Due to the vague and declarative nature of this article, it is also understood that the instructions of the law do not impose a mandatory duty on any of the officers in the company to take into consideration the interests of the stakeholders. This is only a permission granted to the discretion of the company's managers, only if such a consideration is within the business considerations:

> [A]nd within the scope of such [business]considerations, the interests of its creditors, its employees, and the public may be taken into account.

However, even if the article of the Company Goal is drafted differently, in such a way that answers these criticisms, it seems that even then it will not be able to change the managerial culture of the business companies that was established over 200 years ago. The law can never change by itself perceptions or human behaviors. To do so, you need to harness additional social mechanisms, such as the media, social organizations, and of course the system of education, includes the institutions of higher education.

The schools of Business Administration, Economics and Law, are the breeding houses of the future managers, businessmen, directors, accountants and lawyers who shall manage the market in the upcoming years. Therefore, as part of the general perception of COMAS, its different schools and the Academic CSR Center, in order to implant the ideas of Corporate Social Responsibility, it is worthwhile to start educating today the managers of tomorrow.

This all can be done by exposing them to different managerial approaches from Israel and worldwide; by teaching them theories in the twenty-first century in the fields of management, economics and law; by developing a critical sense to the existing theories; by meeting experienced business persons and scholars from Israel and from around the world; and of course by experiencing the market through workshops and activities organized in cooperation with existing businesses.

As stated before, in the Israeli law, the Company Goal article only grants the permission to consider the company's stakeholders, and it does not impose a duty to do so. The fear of the article critics is that as a result, the article shall become ineffective, and the forces of the markets shall affect this possibility not to be exercised. Therefore, the goal that stands before the Academic CSR Center is to make today's students understand the importance of developing a broad perception in all that pertains to the management of companies, and possibly then the permission shall become a custom. A custom according which, the business considerations of the company shall continue to be its ultimate goal, as appropriate for a business and financial entity, and to create a fiscal profit for the shareholders, who invested their financial capital into the company. However, managers should know that a successful and sustainable company is one that acts in harmony with the society and the community, while considering the legitimate expectations of employees investing their human capital, creditors' interests and the public in general, which gives the company the right to exist.

6 Summary

The last decade introduced a change. The access to information revealed questions about the source of power and financial wealth gave birth to the new phenomenon of WEconomy and converting from giving fish to giving fish-hooks. As the leading business and law schools we see ourselves responsible to give our students an up to date knowledge, confront them with reality, and enrich their values, ethics and responsibility. One on the main consequences will be their continuation in creating new business model.

We decided to take the long road and feel fortunate to have open minded, multi-talented personas our SBA dean, as well as a group of dedicated academic staff who is responsible for the re-education of both faculty and students. Additionally, the above changes have created an atmosphere of caring and involvement, enabling young faculty to feel they belong to a mission more important than the daily tasks and responsibilities. The "language" has changed, became more transparent and at the same time clearer and assertive. We look forward to see where the CSR field is developing to, particularly in light of the social changes the world is now witnessing.

Reference

Matten D, Crane A (2005) Corporate citizenship: toward an extended theoretical conceptualization. Acad Manage Rev 30(1):166–179

How CSR Should Understand Digitalization

Andreas Knaut

1 Introduction

There is a lot of euphemism when we talk about the new age of digitalization. We are enthusiastic about the "new digital world". We dream of "Industry 4.0" to raise up efficiency. We emphasize the "Internet of Things (IoT)" to make a 360 degree connectivity come true. We fantasize about "Big Data" and "eCustomer Relationship Management", about the completely transparent customer, about easing delivering processes and having a 24/7 one-to-one-dialogue with the stakeholder.

We create new possibilities of implementing work at home. We succeed to decentralize up to now centralized structures, we theorize of crushing hierarchy and cooperating in an equal way, of giving creativity a virtual room and a chance, of collaborating worldwide.

We romanticize to abolish boring tasks, of giving people the chance to earn their money while working on the beach at sunset.

Whatever the dream is, all the protagonists of the digital future agreed that digitalization means nothing more than turning upside down the way our economy and society have behaved so far. Consequently, they presume a revolution, a so called "disruption" inspired by the German economist Joseph Schumpeter, invented several decades ago to describe a radical destruction of beloved but overaged business models. Disruption became the battle call of the heralds of the digital age.

"Industry 4.0"—as the Federation of German Employers' Associations in the Metal and Electrical Engineering Industries points out—means "a bundle of different projects"—being crosslinked and possessing a number one priority (Gesamtmetall 2016).

A. Knaut (✉)
Knaut Kommunikation, Munich, Germany
e-mail: Andreas.Knaut@yahoo.de

© Springer International Publishing AG 2017
T. Osburg, C. Lohrmann (eds.), *Sustainability in a Digital World*, CSR,
Sustainability, Ethics & Governance, DOI 10.1007/978-3-319-54603-2_21

This message is simple and cannot be over exaggerated. Digitalization is every day everywhere. It is base. It is infecting and will infect everything. It means not only an ever learning IT, which combines data with automatic procedures. Industrial 4.0 combines ever-growing knowledge about the customer with marketing and communication. It defines a total new way of structuring workforces and leading teams. It launches radical change in organizing and structuring a company. It challenges in a fundamental way the manner data will be collected, analyzed and secured.

It is the true and final engine of globalism.

That shutters our understanding of doing projects. It will be no longer from start till end and then shut the light and have a beer. Digitalization means a process of permanent renewing, a culture of perpetual change.

Digitalization is the discotheque to make globalization dance. One depends on another.

In a way, economics has already acknowledged this fundamental architecture of digitalization. Industry 4.0 already introduced the Chief Digital Officer (CDO). The CDO, either a member of the board or a manager at the interfaces of the company, should fix the digital turnaround in a widespread way. His task is to weave the filaments to make the digital carpet fly (Deloitte 2015).

2 We Should Ask Different Questions

Therefore, before companies sightlessly enter the New World, pushing things technically to become more and more efficient—they should take a break and ask themselves some basic questions. To mention some examples:

- What is our vision and mission of digitalization?
- Do we have "digital values"? Do we strive for a hybrid working culture (Ciesielski and Schutz 2016)?
- How would we describe this culture of digitalization?
- Does our sustainable framework reflect a perpetual change?
- What targets we want to achieve in all different areas?
- In one decade, how will we work together, internally and externally?
- What areas will be affected most of all?
- How can we describe the interference of the different changements?
- In which way interface management will be affected?
- In which way job description will be changed?
- Do we have the right people to be prepared and trained for the digital word?
- What should we keep from the analogue culture?
- Does our understanding of sustainability truly reflect the digital world?

Yes, technically everything is possible or will be in some years, but what is really important for our clients?

In addition, yes, transition will be ongoing, but how to organize a culture of permanent change?

More, does this fit to our sustainability strategy? How can digitalization be combined to it?

These questions mean nothing else, than to transfer the analogue culture to a fitting digital culture.

3   Sustainability Ignores Digitalization

So much the disappointment that another buzz of modern industry tends to ignore digitalization. In recognition, that the manufacturing process "is transforming from a patchwork of isolated silos to a nimble and seamless whole fully integrated with the downstream and upstream production environment" (DeAngelis 2016) corporate social responsibility (CSR) ignores this finding.

When we have a look at the recent discussion about a responsible way of doing business or how to organize business workforce to work closely with society, digitalization is not even mentioned. The United Nations Global Contract, one of the most influential tools of modern CSR, does not know a category of digitalization (UN Global Compact 2016). The Global Reporting Initiative lists no category digitalization and asks no specific questions (Global Reporting Initiative 2015).

Of course, digitalization is integrated. Some aspects are reflected in appropriate patterns. In code of conducts we find a lot of remarks about e.g. data safety or how to behave in the social media.

However, CSR deals only with puzzles of digitalization. Digitalization is still regarded as one more technical innovation to be implemented in existing models. This way of consideration will not be enough any more. It does not reflect the whole picture.

4   CSR Needs a Digital Look

CSR models have to give digitalization the importance it needed.

That means that CSR has to reflect its self-conception. We have to understand CSR more as an ongoing process. The nucleus will shift from measurability to processes. The everlasting claim that CSR should become an integrated part of business models will become more actual and more relevant than ever.

CSR should understand and reflect much more the revolution of business models, digitalization really means. To get a better understanding the introduction of new tools might be useful.

A yearly digital transition report, published along or within with the CSR Reports, could be such a tool. Companies should publish their vision and mission of digitalization, reflecting the questions mentioned above. They should show their

Fig. 1 Quadruple
model CSR

stakeholders—especially their employees!—the course of their digital journey and the principles behind.

Such a Digital report considers all aspects of digitalization: production and management processes, data management and security, communication management, human resource, training, measurement, logistics and so on. In addition, in what way different stakeholder will be affected. The digitalization report shows, how the strategy fits in the CSR concept. Relation to sustainability?

Second, I suggest transferring the triple of CSR—Environment, Social, Economic—to a quadruple—Environment, Social, Economic and Digital. Digital will become the fourth column of sustainability (see Fig. 1).

The Digitalization column integrates all aspects of a company's business. It defines the principles of their digital strategy, the targets and the measurement.

It notes e.g. substance rules of the digital way, the way managers and employees collaborate, the way data security is provided and data storehouse is organized. It describes the digital organization. It shows how the company is communicating with the stakeholders. It defines the duties of the digital officer or digital compliance.

It marks KPIs for the development of the digital development—to be followed up in the company's digital report.

5   Aspects of Digitalization

CSR has to rearrange its understanding of optimizing processes. It is not any more to push technology or to collect data. Companies has to change their mindset, to change their view on doing business.

To mention some aspects:

5.1   Share Community

Digitalization means sharing—internally and externally. Leadership culture will change fundamentally. Sustainable leadership in the digital age means sabbaticals, teleworking. Employees will individual manage their working life. They will share their ideas, their working place, they will portion their working time. In addition, workforce training will become more permanent (Thom and Zaugg 2002).

In the digital Age we develop hybrid working rooms, we will work in ensembles on common platforms. Management and Employees will learn to collaborate, to share knowledge and ideas on common digital platforms. They will do it an open and free-minded way, without the fear of being blamed for a bad idea or that colleagues might catch a suggestion to highlight their own position. Groups with different members will be virtually formed and equally dismantled. They communicate globally; they join forces tactically. Speed matters.

Externally we learn to drop our thinking of competition. Competitors in one market could be partners in another. They even can be collaborators in the same market in 1 minute and competitors in the next. Companies have to evaluate their position in the market every minute.

Customers and clients become partners, even ambassadors and participants of the brand. Marketing and Communication will integrate; the social media department becomes the center of client relation. In particular, we will observe that industries doing business to business will revolution their client dialogue. Digitalization provides the possibility to create a 24/7 link to the customer. Offline and online communication will arrange their ways in a different manner.

Communicating with a client means not only selling a product but to build up a relationship. It means that companies have to share knowledge and detail information, e.g. the DNA of the brand. The client will become a well-informed friend, a partner who will influence the way producing and creating the product. "After sales" will get a different meaning. Partnership means as well, that from time to time also disagreements might come up. Companies should reflect criticism as a chance not a threat.

In a nutshell the employee of the digital age will follow his working life much more self-determined than his colleague is doing today. And he has to develop much more skills of self-organization.

What does that mean to CSR? The standardizing body of sustainability has to reflect these developments. What is our vision of a hybrid working place? Does our company push these development? Do we have a suitable leadership culture and an appropriate working environment? How can we describe the values we lead, we work together and we communicate to each other?

5.2   Lost Control

Future Management has to accept that it will lose control. So far internal communication provide issues and channels to garden the employees. However, it will soon state, that the old kingdom of mass manipulation will finally crack down. Social media, just-in-time media, multichannel communication and the need of much more internal collaboration than before provoke a quantum jump of interpersonal exchange which cannot be controlled any more.

This has all begun. Have a look at platforms like yammer. We diagnose an internal chat, where employees emancipate from the guided official way of internal platforms and find new ways to organize themselves. Not a bad way to encourage the acceptance of your company.

In the outside, we observe the same development. The old model of monologue marketing is passé. Right now we begin to understand what a 24/7 dialogue with stakeholders really means. Corporate brands have to explain themselves permanently. Mass marketing, e.g. in TV, is still needed to launch the brand, but cannot channel reactions any more.

Therefore, we have to create a new idea of reputation management. CSR will no longer be a sidekick of corporate communication; it will move to the center of issue management. CSR has to define values and arguments, which reflects attitude and responsibility of the company. CSR need to build criteria and guidelines for a reliable communication. In the nearby future reputation management will be fundamentally based on a sustainable story.

5.3   Permanent Transition

Permanent transition is not only a buzz to shutter computer nerds or to highlight the yearend speech of the CEO.

For most of the employees—and customers as well—a world of permanent change means an ongoing threat, a dark cloud of insecureness in a limitless world. The world of tomorrow is no more reliable, easy to understand and straightforward. Instead we gain a volatile world, complex, contradictory, unstable.

Therefore, permanent change describes a different corporate culture. Management and employees have to accept transition as a vital and ongoing part of their daily work. They have to be ready to embrace it as a change for creativity and forthcoming not as a threat to be fired. Mobility will be the wiz of the future.

However, everybody gets a chance to be part of the future and to create it in his workplace for the sake of the whole.

Digitalization means as well, that companies will lose control of their own digital development. The process of digitalization will spread so rapidly and will affect business in such complexity, that companies have to accept that they are not able to predict every outcome of the process. It may sound paradox, but this finding should be included in planning.

As well sustainable measurement and planning has to accept that phenomenon. Digital development will become more cross-linked and more unstable and less predictable than ever. We already have difficulties to forecast and to simulate complex correlations in global business, we will have more in the future (Lexikon der Nachhaltigkeit 2015). For that reason, we have to discuss what sustainable development in the future really means.

5.4 Leadership from the Top

Management by chaos? Are we in a small boat alone in the ocean with no clue, where the stream is heading for? That is true. This exactly means digitalization.

That is why companies need leadership. They need to develop a vision of digitalization and to give stakeholders an idea of it. They need to lead from the top. To give orientation to the oarsmen and to provide security, when whales shutter the wooden ship or the stream surprisingly turns its direction. One finding of a recent study of Deloitte (Deloitte 2015).

That's why a true sustainable management should give trust to employees and external stakeholders in the way the company manages digitalization. It should explain, how innovation and implements will be implanted. It should give people an impression how management will decide digital innovation.

Digitalization means to think and to manage process from the outcome. We have to transfer our analog thinking in a modern digital culture and to develop "digital values". CSR has to play a vital role in that discussion.

6 Conclusion

Corporate Social Responsibility is able to provide a sustainable leadership. Assumed, it has accepted digitalization as base of the modern world, not only as technical phenomenon. For 20 years, digitalization has revolutionized the way people organize themselves, work, and communicate.

Therefore, we urgently have to define what does that mean to the concept of sustainability and CSR.

CSR has to reconsider its self-concept. It has to shift its focus from measurement to processes. It has to adopt and to implement the multi complex, the multi linked way of the digital and global business world. And it has to deal with it in high-speed.

Companies have to define their digital culture. They should understand digitalization as a holistic and hybrid driver of their business model.

A yearly digital report could be helpful to systematize this process. A fourth column model of CSR helps to manage it.

We have to think about digitalization from the end. We have to accept its nature and then to create it.

References

Ciesielski MA, Schutz T (2016) Digitale Führung: Wie die neuen Technologien unsere Zusammenarbeit wertvoller machen. Berlin

DeAngelis SF (2016) The Internet of things and industry 4.0. http://supplychainminded.com/internet-industry-4-0

Deloitte (2015) Studie Überlebensstrategie Digital Leadership, München. www2.deloitte.com/de/de/pages/technology/articles/survival-through-digital-leadership.html

Gesamtmetall (2016) Industrie 4.0, Berlin. www.gesamtmetall.de/themen/beschaeftigung/werkvertraege

Global Reporting Initiative (2015) G4 sustainability reporting guidelines. New York. www.globalreporting.org/resourcelibrary/GRIG4-Part1-Reporting-Principles-and-Standard-Disclosures.pdf

Lexikon der Nachhaltigkeit (2015) www.nachhaltigkeit.info/artikel/ist_100_prozent_nachhaltigkeit_moeglich_1781

Thom N, Zaugg RJ (2002) www.researchgate.net/profile/Norbert_Thom/publication/269709349_Das_Prinzip_Nachhaltigkeit_im_Personalmanagement_Ausgewhlte_Ergebnisse_einer_Befragung_in_europischen_Unternehmen_und_Institutionen/links/54947e280cf29b944820616f.pdf

UN Global Compact (2016) Making global goals local Business, New York. www.unglobalcompact.org

Sustainable Development Goals

Goal 1:   End poverty in all its forms everywhere
Goal 2:   End hunger, achieve food security and improved nutrition and promote sustainable agriculture
Goal 3:   Ensure healthy lives and promote well-being for all at all ages
Goal 4:   Ensure inclusive and equitable quality education and promote lifelong learning opportunities for all
Goal 5:   Achieve gender equality and empower all women and girls
Goal 6:   Ensure availability and sustainable management of water and sanitation for all
Goal 7:   Ensure access to affordable, reliable, sustainable and modern energy for all
Goal 8:   Promote sustained, inclusive and sustainable economic growth, full and productive employment and decent work for all
Goal 9:   Build resilient infrastructure, promote inclusive and sustainable industrialization and foster innovation
Goal 10:   Reduce inequality within and among countries
Goal 11:   Make cities and human settlements inclusive, safe, resilient and sustainable
Goal 12:   Ensure sustainable consumption and production patterns
Goal 13:   Take urgent action to combat climate change and its impacts*
Goal 14:   Conserve and sustainably use the oceans, seas and marine resources for sustainable development
Goal 15:   Protect, restore and promote sustainable use of terrestrial ecosystems, sustainably manage forests, combat desertification, and halt and reverse land degradation and halt biodiversity loss
Goal 16:   Promote peaceful and inclusive societies for sustainable development, provide access to justice for all and build effective, accountable and inclusive institutions at all levels
Goal 17:   Strengthen the means of implementation and revitalize the global partnership for sustainable development

About the Authors

Jan Philipp Albrecht is spokesperson for Justice and Home Affairs of the Greens/EFA in the European Parliament. He is the vice chair of the Committee on Civil Liberties, Justice and Home Affairs (LIBE) and a substitute member of the Committee on the Internal Market and Consumer Protection (IMCO). During his first mandate between 2009 and 2014 he was a member of the Committee on Civil Liberties, Justice and Home Affairs (LIBE) and a substitute member of the Committee on Legal Affairs (JURI). From December 2012 to October 2013 he was the Green group's coordinator in the Special Committee on Organised Crime, Corruption and Money Laundering (CRIM). He was rapporteur for the opinion of the Committee on Legal Affairs on the proposal for a directive on the right of access to a lawyer in 2011. Since March 2012, Jan Philipp Albrecht is the rapporteur of the European Parliament for the data protection regulation. From 2003 until his election to the European Parliament in 2009, Jan Philipp Albrecht studied law in Bremen, Berlin and Brussels and specialised in IT law at the Universities of Hanover and Oslo. Since 1999 Jan Philipp Albrecht has committed himself to the Greens in a wide range of contexts. Thanks to his efforts to promote data protection, the former federal spokesman of the Young Greens in Germany (2006–2008) has rapidly gained a reputation within the European Parliament as an expert on home affairs, justice and legal affairs. Jan Philipp Albrecht was born on 20 December 1982 in Braunschweig.

Sezen Aksin-Sivrikaya holds a Bachelor of Science degree in Industrial Engineering from Koc University, Istanbul, Turkey and a Master of Science degree in Economics and Management Science from Humboldt University, Berlin, Germany. She is currently a doctoral candidate at Humboldt University, Berlin, Germany. She is also the Research and Teaching Coordinator at the Center for Sustainable Business at ESMT Berlin. She has been teaching at a number of institutions including Humboldt University, ESMT Berlin, and BAU International Berlin. Her research focuses on sustainability, corporate responsibility, corporate governance, and corporate reputation.

© Springer International Publishing AG 2017
T. Osburg, C. Lohrmann (eds.), *Sustainability in a Digital World*, CSR, Sustainability, Ethics & Governance, DOI 10.1007/978-3-319-54603-2

CB Bhattacharya is the Pietro Ferrero Chair in Sustainability and Director of Center for Sustainable Business at ESMT European School of Management and Technology in Berlin, Germany. Prof. Bhattacharya has published over 100 articles and has over 15,000 citations per Google Scholar. He is co-author of the book *Leveraging Corporate Responsibility: The Stakeholder Route to Maximizing Business and Social Value*. He places 10th in the category Top 100 current researchers and 14th in the category Top 250 researchers—lifetime work in the Handelsblatt Business Administration Ranking. He has consulted for many organizations such as Allianz, AT&T, Eli Lilly, E.ON, Procter & Gamble Company, and others. He is often interviewed and quoted in publications such as *Business Week, Forbes, Financial Times, Newsweek, The New York Times* and *The Economist*. He received his Ph.D. in Marketing from the Wharton School, University of Pennsylvania in 1993 and his MBA from the Indian Institute of Management in 1984.

Görkem Çokçetin has been working in Turkish banking sector since 1998, just after his graduation from University. He spend 14 year in Corporate Banking with different roles in network and head office starting from RM to Regional sales manager. He worked in Garanti Bank and Finansbank and joined Yapı Kredi (a member of Unicredit) in 2005. In 2012 he was appointed as Idea Management Center Manager to establish the unit and internal ideation and innovation system for employee called Evreka. Evreka system is receiving around 9000 ideas in each year with an increasing trend. Evreka is being perceived as a best practice for idea management in Unicredit group and Turkish banking sector. Evreka won an award from CSR Turkey in Corporate Entrepreneurship category in 2015. He got his master degree in Banking and Finance from Bilgi University with his thesis on "Mobile Banking and It Effects in Banking Sector with 3 a Year Projection". He is a member of Koç Holding (the biggest industrial group of Turkey) Innovation and Technology Council and TUSIAD (Turkish Industry and Business Association) Innovation Work Group. He is giving trainings on "Innovation and Innovation in Finance Sector" in Yapı Kredi Banking Academy and in programs conduct with various Turkish Universities.

Thomas F. Dapp has worked since 2008 as an Economist at Deutsche Bank Research, the independent think tank of Deutsche Bank AG in Frankfurt am Main. In a broad sense his research focuses on topics such as innovation, digital structural change and the digital economy in general. As an associate he worked on a one-year research project for the "Stiftung Neue Verantwortung" in Berlin which investigated "the power of innovation in digital ecosystems". His most recent publications are in the areas of mobile payment, Big Data, FinTech and crowdfunding/crowdinvesting.

Amira Dotan, Brig. General (Res.), former MK. Founder and Joint CEO of the Mediation Center, Neve-Tzedek. Chairperson of the Israeli Academic Research Institute of Conflict Resolution and Mediation, The College of Management Academic Studies. Chairperson of the Center for Corporate Social Responsibility—School of Business Management—The College of Management Academic Studies. Served as a board member at various Israeli business companies. Professional Activities: I.D.F.—Chief of the Women Corps Ben-Gurion University—Vice

President. Chairperson of the Zionist Delegation and the representative of the Jewish Agency Executive in North America.. CEO, Operation Independence. Received the awards of International Advocate for Peace Award (2009)—Cardozo School of Law, Yeshiva University.

Tobias Engelsleben is Professor for Business at Fresenius Hochschule. Being a Marketing and Strategy Consultant he started his academic career in 2003. Tobias joined Fresenius Hochschule, in 2005. Fresenius Hochschule is a private University in Germany, founded in 1848 with today 10,000 students. Since 2016 Tobias is President of Fresenius Hochschule.

Anat Even-Chen is the Manager of Academic CSR Center, at the School of Business Administration, the College of Management Academic Studies (COMAS), Rishon LeZion, Israel. She received her MBA from the School of Business Administration, COMAS, Rishon LeZion, and her Bachelor degree in Communication and Far East Studies (chosen language Japanese), from Tel-Aviv University, Israel. Her current research focuses on- "The effect of the Perceived Image of a Social Business on its Consumers' Experience."

Denise Feldner, Ass. jur., founding managing director of U15, is a graduate lawyer who holds an executive master's degree in business law for tech companies. She has been working as chief of staff to the rector of Heidelberg University for four years. Prior to this role she served on behalf of the university's rectorate as observer in a university- and industry-owned start-up for research in organic electronics. She worked as in-house counsel for the company's CEO. Before that, she held a variety of lawyer positions at the German Federal Ministry for the Environment, at the Higher Regional Court of Berlin, and in an international corporate law firm. Her law studies took her to Berlin, Leuven, Budapest, and Athens. In Budapest she joined ELTE Bibó István College, Legal and Social Sciences Institute for Advanced Studies. Feldner is a Helmholtz Academy fellow and certified manager by Malik Management St. Gallen. Before finishing her law studies she has been working for eight years as head of research team with a law firm. She has been working as registered legal clerk for four years. For her apprenticeship she worked with a notary's office. Feldner was named 'Young Leader' by the Confederation of German Industry (BDI), by Wilton Park, British German Forum, and recently by the German Russian Young Leaders Conference. The Weizman Institute of Science, Israel, invited her to become member of the Weizman Young European Network.

Christiane Gebhardt has been with the Malik Management Zentrum St. Gallen/Switzerland for 17 years and currently holds the position of Vice President/Head Global Initiatives. Christiane was a member of the High Performance Cluster Commission for the German Government from 2009 to 2014 and took part in the expert group Industry 4.0/Smart Service World at the German Academy for Science and Engineering. She is an expert for Chinese SAFEA and European DG Regional & Urban Growth. She has evaluated Chinese, European and US innovation programs and carries out consulting for larger firms, research organizations, clusters and scientific start-ups. Christiane holds a master degree in Public Administration Science from University of Konstanz and obtained her Ph.D. from Giessen

University in Political Sciences for her thesis on Regionalisation of R&D Policies in the US and Germany. Her scientific career includes several stations: Jena University, State University of New York and research stays at Bristol University, UMASS at Amherst, Chicago University and MIT Boston. Her research is on regional innovation systems linked to Strategy, Governance and Organizational Design. Christiane teaches at Heidelberg University and is an active member of the Triple Helix Association.

Carl-Otto Gensch is Head of the Sustainable Products & Material Flows Division at Öko-Institut e.V. in Freiburg. He was intensively involved in the practical and methological development of Life Cycle Assessment, including the harmonization at ISO level. Furthermore he has long-lasting experiences conducting life cycle cost analysis (LCC) and eco-efficiency analysis. With his extensive background on (environmental) impact assessment he is well experienced to analyse at a more general level technological pathways of innovation as well as specific technologies, including broad experiences regarding market and data analysis. Carl-Otto Gensch has good and long-lasting contacts to experts working in the field of technological innovation of several sectors. As process engineer he managed a large number of research projects dealing with the compilation, analysis and assessment of complex systems. Based on his long-standing methodological know-how (LCA, LCC, eco-efficiency analysis, Product Sustainability Assessment PROSA) he is expert in robust assessments of the environmental impacts and benefits of products, processes and systems. Currently he is the project leader of a re-search project "Trafo 3.0" The overall objective of this project is the development and testing of a heuristic, and the drafting of a manual planned as an E-Book to support practitioners in contributing to the initiation and in actively shaping socio-ecological transformation processes.

Zeynep Gürhan-Canlı is Migros Professor of Marketing at Koç University, Istanbul, Turkey. She completed her Ph.D. in marketing at New York University Stern School of Business in 1997. Prior to joining Koç University, she was a tenured faculty member at Ross School of Business, University of Michigan. Her research interests include consumer information processing in relation to branding, corporate image and corporate social responsibility. Her recent research with UNDP IICPSD focuses on the role of private sector in development. She has published several articles in leading academic journals such as *Journal of Consumer Research, Journal of Consumer Psychology,* and *Journal of Marketing Research.* She is an associate editor for the *Journal of Consumer Research* and senior editor for the *International Journal of Research in Marketing.* She is on the Editorial Review Boards of the *Journal of Consumer Psychology, Journal of International Marketing, Journal of the Academy of Marketing Science, and Journal of Marketing Behavior.*

Hans-Dietrich Haasis studied industrial engineering at the University of Karlsruhe from 1978 to 1983. After his doctorate in 1987 and his habilitation in 1993, he also became a university professor of business administration at the University of Bremen. Since 1997 he has been General Secretary of Business Administration, Production Management and Industrial Management at the University of Bremen. Since 2015, he has been renamed to General Business Administration, Maritime Economics and Logistics. From July 1998 until June 2001, he was first speaker, then Dean of the Department of Economics. He was a guest professor at the Ecole Nationale Superieure du Petrole et des Moteurs, Paris, Rueil-Malmaison, at the Catholic University of Eichstätt-Ingolstadt, at the University of Witten-Herdecke and at St. Petersburg State University. In 2003 he received the B.A.U.M environmental award. He is a founding member and was spokesman of the research group Logistics of the University of Bremen from July 2002 to June 2005. From August 2001 to December 2014, he was Director and Head of the Logistic Systems Division, and from January 2007 to December 2014, Chairman of the Board of the ISL—Institute for Shipping and Logistics (www.isl.org), Bremen. Since October 2012, he has been the spokesman for the IGS—International Graduate School for Dynamics in Logistics at the University of Bremen. He is the author of numerous publications and is a member of several scientific advisory boards and expert committees. In particular, he is a member of the Scientific Advisory Board of the Federal Minister for Transport and Digital Infrastructure, member of the Board of Trustees of the BVL-Campus of the Federal Association of Logistics, EMA, Euro-Mediterranean-Arab Association and spokesman for AGKN, Asian-German Knowledge Network of Transport and Logistics.

Markus Hartwig is Power Trading Expert and currently works for Smart Energy, Hitachi Consulting.

Olaf Heil is responsible for Hitachi's ICT and OT solutions for these areas in the EMEA-CIS region. Currently, the platform-based software solutions are of great interest to cities and energy suppliers. Dr. Heil has a 25-year management experience in power generation and marketing, as well as in the development of new technologies and innovation management. He previously worked as a division manager and managing director of the RWE Group and has successfully completed numerous energy-related projects in Germany and abroad.

Inga Hilbert works as a scientific assistant in the Sustainable Products & Material Flows division at Öko-Institut in Freiburg. Her work is focused on socio-economic transformations, as for example the Energiewende, and the question, how these processes can be stirred in a more sustainable direc-tion. Additionally she works on the impacts of increasing digitalization and the question, which alter-natives for a more sustainable implementation exist. In the course of her work she assesses different technologies and their impact on societal processes. Before she started working at the Öko-Institut in 2015, she worked as a student assistant in energy consulting and did an internship at the Interna-tional Climate Initiative (IKI) in Berlin. Inga Hilbert holds a Bachelor degree in Industrial Engineering.

Alexander Holst is Managing Director at Accenture Strategy and leads the Sustainability Practice in Germany, Austria and Switzerland. He focuses on integrating sustainability aspects into the core business of clients across numerous industries. Alexander has led a broad range of strategy projects for corporate clients, looking at how to value environmental and societal aspects and how to capture financial benefits by integrating these aspects in product portfolios, processes, and business models. He is a frequent conference speaker and has published on various sustainability topics.

Alexander studied European Business in Osnabrück and holds an MBA from IESE, Barcelona. He is a member of the "Business & Sustainability Program" at the University of Cambridge. Alexander works out of Accenture's Berlin office.

Andreas Jung is Member of the German Bundestag for CDU/CSU group. Andreas, himself a lawyer, is specialized on nature and sustainability issues. Among other positions he is Chairman of the Parliamentary Advisory Council on Sustainable Development, Member of the Committee for Environment, Nature Protection and Nuclear Safety and Commissioner of the CDU/CSU Bundestag group for climate protection as well as Member of the Committee for Economy and Technology.

Katharina Klug is Professor of Marketing at Fresenius Hochschule, Akademie für Mode und Design in Munich. Previously she was responsible for Market Research at AOK PLUS in Dresden und worked as a research fellow at TU Dresden. She studied Economics at TU Dresden, Università degli Studi di Trento (Italien) got her Ph.D. from Christian-Albrechts-Universität zu Kiel iin 2014. She is specialized in innovative forms of communication as well as consumer behavior and intercultural marketing.

Andreas Knaut is consultant and interim manager for corporate communication and sustainability. He lectures at Fresenius University, Munich, and is chief editor of demografiewandel.info, first German online medium for demographic change. For 25 years Knaut has worked as director corporate communication and CSR for Danone, SCHUFA Holding, Gruner+ Jahr and Verlagsgruppe Handelsblatt. He was chief editor of "Der Kontakter" and deputy chief editor of "werben und verkaufen".

Carl-August Graf von Kospoth is Member of the Board of BMW Stiftung. The BMW foundation aims to connect leaders from different sectors, cultures and nations to promote social innovation, global dialogue and responsible action for sustainable development. Carl-August has an academic background in Engineering and worked in the mobility business in Munich, Berlin and Brussels.

Martin Kreeb is Professor and study dean for Sustainable Marketing and Leadership as well as study dean for the General Management (MBA) Program at the Fresenius Business School in Munich. Martin studied at the Universities of Hohenheim, St. Gallen and Witten/Herdecke. He is specialized on Environmental and Energy Management, Logistics as well as Sustainability Management and Communication.

In 1999 he founded the German Competence Center for Sustainable Management at the University of Witten/Herdecke. From 2000 to 2016, he led the Research Group on Sustainability at the Chair of Environmental Management at the University of Hohenheim. Since 2017 he has been Research Fellow at the University of Bremen in the field of Green Logistics and Sustainability at the Chair in Maritime Business and Logistics of Hans-Dietrich Haasis.

Gerd Leonhard is a Futurist and Author, Keynote Speaker, Think-Tank Leader & Advisor. Gerd also serves as Visiting Professor at the Fundação Dom Cabral (São Paulo, Brazil). Gerd's background is in music and the music business in the 1980s. He then caught the Internet bug and became a digital media entrepreneur and start-up Internet CEO. He speaks at many major conferences and events, company retreats, seminars and in-house trainings on the Future of Business, Sustainability, Environmental Policy and Climate Change, Media, Content and Entertainment. In 2006, The Wall Street Journal called Gerd 'one of the leading Media Futurists in the World'. Gerd is a fellow of the Royal Society for the Arts (London) and a member of the World Future Society. A native German and long-time US-resident (San Francisco and Boston), he now resides in Basel, Switzerland.

Christoph R. Löffler is Group Director for Fjord in Austria, Switzerland and Germany. Christoph firmly believes we are living in one of the most exciting, transformative but also demanding periods in modern history and we have an incredible privilege to be active architects of change. With an academic background in analytics, Christoph began his professional career in the USA. Before converting entirely to the service design world, he took the challenge from Accenture Interactive to build up a new generation of digital agency for the Greater China region. When Accenture Interactive acquired Fjord globally in 2013 he moved from Shanghai to Berlin to lead Fjord's integration across Germany, Austria and Switzerland.

Christiane Lohrmann is head of Marketing and PR at FranklinCovey Leadership Institut in Germany, Austria and Switzerland, an international consulting company specialized on Company Culture, HR and Organisational Development. She lectures at Ludwig Maximilian University and Fresenius University in Munich about Sustainability Management and Communications as well as Business Ethics. For 20 years Christiane was responsible for Marketing Strategy and PR for FOCUS Newsmagazine with Hubert Burda Media. Her espertise is in Marketing, Strategy, CSR, Education and Learning. She holds an MBA Sustainability and CSR from Leuphana University Lüneburg and an M.A. in American Studies, Communications and Political Science from Ludwig Maximilian University Munich.

Ivo Matser was boardroom consultant in many companies in various industries over the past 14 years and had several senior management/CEO positions in business. From 2003 till 2014 he was CEO of TSM Business School and in the last 2 years he is CEO of ISM University of Management and Economics and dean of the ISM's Executive School. Dr. Ivo Matser is board member of several companies and in addition to that he is involved in associations as EFMD, The Academy of Management (US), Global Peter Drucker Forum, Principles of Responsible Management Education of the United Nations (PRME and GRLI), European Leadership Platform and international accreditation bodies for higher education. Also he is one of the European Marketing Professionals, EMP, doctorate level based on experience. Dr. Ivo Matser holds a Doctor Honoris Causa (Dr. hon) title because of his efforts higher education to become more entrepreneurial and relevant.

Seda Müftügil Yalçın is a post-doctoral research fellow and instructor at Koç Univerisity Social Impact Forum. After graduating from Sabancı University (SU), she completed her masters degree at London School of Economics (LSE) in Human Rights. Later, she obtained her Ph.D. from University of Amsterdam, Amsterdam School for Cultural Analysis (ASCA) in the field of Cultural Analysis. Her current research focuses on inclusive businessess and practices; social impact management and its measurement; social enterprises and social entrepreneurship.

Thomas Osburg is Professor for Sustainable Marketing & Leadership at the Fresenius Business School in Munich and Director of the CircularKnowledge Institute, an International Research ThinkTank and Strategic Advisory. For more than 25 years, Thomas worked for global IT companies (Intel, Autodesk and Texas Instruments), living in France, the U.S. and Germany. He is on the Board of Directors for ABIS (Academy of Business in Society, Brussels) and was appointed into various Scientific MBA Committees at leading European Universities, where he is frequently teaching MBA classes on Technology and Innovation Management, Strategic Marketing, Social Innovation, Entrepreneurship and CSR. He has published several books on Social Innovation and CSR Marketing and written over 30 scientific contributions for leading European journals.

Sebastian Philipps is consultant at Accenture Strategy—Sustainability in Berlin. He has been working on sustainability projects since 2007 with particular interest in complicated stakeholder settings that involve public-private interfaces and the consumer. Prior to joining Accenture Strategy, Sebastian was responsible for the corporate partnerships strategy at the Asian Development Bank's Core Environment Program in the Mekong region. At the UNEP/Wuppertal Institute Collaborating Centre on Sustainable Consumption and Production he managed various China projects and developed the portfolio around sustainable finance. Sebastian is an economist and China scholar by training, with an emphasis on behavioral facets of business and society.

Siddharth Prakash is Senior Researcher at the division "Sustainable Products and Material Flows" of the Öko-Institut e.V. in Freiburg. He is an expert in analysing the sustainability impacts of infor-mation and communication technologies. He is the project leader of several R&D projects conducted on behalf of the Federal Environment Agency of Germany in the thematic area of Life Cycle Assessment of ICT. In this regard, he recently completed a milestone study on "Strategies against Obsolescence" with focus on electrical and electronic devices. Furthermore, he has developed the Sustainable Assessment Framework (SASF), a comprehensive tool to assess the sustainability performance of products and services of the global ICT industry, for the Global e-Sustainability Initiative (GeSI). Apart from that, he has been responsible for developing the label criteria for the German eco-label, the Blue Angel, for a large range of ICT products, such as green data centres, mobile and desktop computers, monitors, DVD/Blu-ray products, HiFi-systems etc. In the EU eco-design process, he advises the European consumer organisations BEUC (Bureau Européen des Unions de Consommateurs) for ICT-products. In the past, he had also advised the European Commission in the field of methodologies for measuring and reporting on greenhouse gas emissions of ICT products, services and companies. In the field of e-waste management, he has worked on behalf of the United Nations Environment Program (UNEP), the Ministry of Environment of the Netherlands (VROM) and German Development Bank (KfW) on the development of sustainable global closed material loops in West Africa. Currently, he is working on the sustainability issues related to the ongoing digitalization process, internet of things and Industry 4.0.

Yossi Rahamim is associate professor at the Haim Striks School of Law, the College of Management Academic Studies (COMAS), Israel. He is an expert on Labour Law, Industrial Relations and Corporate Law. He received his Ph.D. from Bar-Ilan University; LL.M. from Tel-Aviv University; and LL.B. from Haim Striks School of Law, the College of Management Academic Studies. He is the author of the book "Employees' Rights in the Event of Transfers of Undertakings". His current research focuses on "Corporate Law as a Stakeholders Protection Device".

Stephan Rammler studied Political Science and Economy. He got his Ph.D. from Wissenschaftszentrum Berlin for Social Science (WZB). Since 2002 he is Professor at Hochschule für Bildende Künste in Braunschweig and since 2007 founding Director of the Institute for Transportation Design. His research topics are Mobility, Future, Traffic-, Energy and Innovation, as well as cultural Transformation and future Environment and Social subjects. He holds the ZEIT WISSEN Award for Sustainability.

Alfons Sauquet Rovira is the Global Dean of ESADE Business and Law School since 2014 and currently Chairman of the Board of Directors at ABIS. He previously served as Dean of ESADE Business School from 2007 to 2014 and as Dean of Research and Director of University Programs. Sauquet holds a number of esteemed positions with other networks including European Foundation for Management Development (EFMD), Global Management Admission Council (GMAC) and has been invited to speak at numerous international fora. Sauquet holds a Doctorate from Columbia University, Master in Org. Psych. from Columbia University, MBA from ESADE; Degrees in Philosophy and Psychology by the University of Barcelona.

Stefan Schepers is Secretary General of the tripartite High Level Group on Innovation Policy Management in EU (2012-). Being a visiting professor European Studies, Henley Business School, University of Reading (UK) Stefan is Consultant on European policy and on corporate strategy, EPPA and AEAC. He was former Director general of the European Institute of Public Administration (Maastricht, 1981–1990). He is Member of the Senate of the European Academy of Sciences and Arts, Chairman of the Governing Council of Mazungumzo—the African Forum in Brussels (www. mazungumzo.eu). Stefan holds a Master in Law, University of Leuven (B), Master in Advanced European studies, University of Strasbourg (F), Ph.D. in Political Sciences, University of Edinburgh (UK). He published articles on EU governance and corporate management. Among his recent books are: Rethinking the future of Europe (2014), co-ed. Andrew Kakabadse. Revolutionising EU Innovation Policy, co-ed. Klaus Gretschmann (2016). Both at Palgrave MacMillan.

Franz Wenzel is researcher, lecturer and entrepreneur. He studied Business Economics and Management with the major subjects Marketing, Informatics, Psychology and Adult Education at Catholic University/Ingolstadt School of Management. His research in the domains of Innovation and Management both on Small and Medium-Sized Enterprises as well as on Start-Ups form the center of his academic work. Holding expertise in Management Strategy, Leadership and Socio Economics (viewed from the perspective of Sustainable Development), Franz developed an innovative Megatrend approach and his concept of Cooperative Innovation. He is entrepreneur (owner of a real estate business with over 25 years of own experience) and self-employed in the fields of education, media and digitization. Franz is executive officer of an international eCommerce company, managing Director of a research institute, scientific Director of an innovation think tank and Start-Up investor.

Dicle Yurdakul is Assistant Professor of Marketing at İstanbul Kemerburgaz University. She earned her Ph.D. in Business Administration (with a marketing major) from Izmir University of Economics in June 2013. She served as a lecturer at Izmir University of Economics and taught marketing management, marketing research and retailing at graduate and post-graduate levels. She conducted her post-doctoral research at Koç University between February 2014 and September 2015. She is also performing as a consultant for UNDP IICPSD projects on the role of private sector in development. Her research interests lie in the fields of poverty research, consumer empowerment and sustainable development.

The manufacturer's authorised representative in the EU is Springer
Nature Customer Service Centre GmbH, Europaplatz 3, 69115 Heidelberg,
Germany. If you have any concerns regarding our products, please
contact ProductSafety@springernature.com

Printed and bound by CPI Group (UK) Ltd, Croydon, CR0 4YY
29/04/2026
02099450-0018